ASSESSMENT BOOK

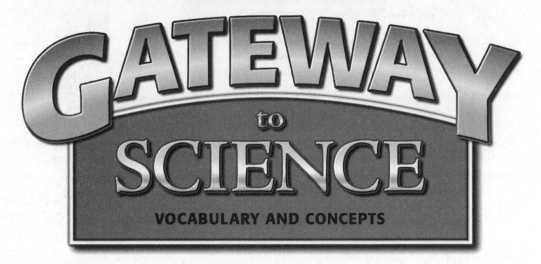

GATEWAY to SCIENCE

VOCABULARY AND CONCEPTS

Tim Collins
Mary Jane Maples

THOMSON

HEINLE

Australia • Canaada • Mexico • Singapore • United Kingdom • United States

THOMSON
HEINLE

Assessment Book
Gateway to Science
Mary Jane Maples

Publisher, School ESL: *Sherrise Roehr*
Director of Product Development: *Anita Raducanu*
Senior Development Editor: *Guy de Villiers*
Director of Product Marketing: *Amy Mabley*
Executive Marketing Manager: *Jim McDonough*
Associate Production Editor: *John Sarantakis*

Print Buyer: *Susan Carroll*
Composition: *Parkwood Composition Service*
Cover design: *Chrome Media/InContext Publishing Partners*
Printer: *West Group*

Printed in the United States of America.
1 2 3 4 5 6 7 8 9 10 — 11 10 09 08 07

For more information contact Thomson Heinle, 25 Thomson Place, Boston, Massachusetts 02210 USA, or you can visit our Internet site at http://elt.thomson.com

ISBN-10: 1-4240-0894-8
ISBN-13: 978-1-4240-0894-0

Contents

Assessing English Language Learners in Science

Appropriate assessment of English language learners in science involves using appropriate formative, summative, and alternative assessment. All assessment of English language learners should be adapted to the learners' language level in order to assess subject matter knowledge and not language ability. For this reason, alternative assessment, including drawing or responding to pictures, responding orally to questions, or hands-on projects, valuable for all learners, may be particularly useful with learners with lower proficiency levels. In addition to assessing learners appropriately, such assessments build learner confidence as their language skills increase.

Formative Assessment

Use formative assessment at the beginning of each lesson to gauge learners' prior knowledge. You may use the ExamView® test generator on the Gateway to Science Teacher Resource CD-ROM to create an appropriate pre-assessment, or you may assess learners' prior knowledge through prompts and questioning. You might show learners a relevant picture or have them look at the first spread of the lesson. For some lessons, you might bring appropriate realia, such as lab equipment, rocks and minerals, or simple machines. Then encourage the students to say as much as they can, allowing you to pre-assess their knowledge of the target language and concepts. Then adjust your teaching objectives accordingly.

During each lesson, use ongoing, informal assessment to monitor students' progress. Each day, focus on two or three students, and note down areas of strength and areas that need further reinforcement or development. Then provide relevant re-teaching or review. You might use a special journal, notebook, or computer file to record your observations.

Summative Assessment

Use the assessments in this Assessment Book to assess learners at the end of each lesson, at the end of each section (Science Basics, Life Science, Earth Science, and Physical Science), and upon completion of the book. Use the ExamView® test generator on the Gateway to Science Teacher Resource CD-ROM to create custom assessments for each lesson. In order to adapt traditional assessments to the language level of your students, consider any of these suggestions:

- Allow learners extra time to complete the assessment.
- Read or have an aide read the assessment aloud in English to the student.
- Read aloud or have a bilingual aide read the assessment aloud in the student's native language.
- Allow learners to use a bilingual dictionary while completing the assessment.
- Reduce the number of options in multiple choice items by reducing the number of distractors (incorrect answer options) by one.

In addition, you may use any of the alternative assessment suggestions below in order to adapt assessment to students' language level, or to gather additional assessment data about all learners.

Alternative Assessment

Alternative assessment includes many non-traditional ways to assess learner projects, and they are ideal for assessing learners of lower language levels. Moreover, alternative assessments are ideal for providing rich, motivating assessments for all learners. Alternative assessment techniques include:

- Labeling pictures or diagrams. For example, students might label a picture of a plant with the names of its parts (flower, stem, etc.).
- Drawing and labeling pictures or diagrams. For example, students might draw a diagram of the layers of the atmosphere and then label each level with its name.
- Including drawings or diagrams in lab manuals or reports. For example, in a simple lab in which students observe what happens when they place a candle in a jar, they can create labeled pictures of the steps they followed instead of describing the steps and the outcome in words.
- Collections or samples. For example, in a lesson in life sciences, students might take pictures of the leaves of various kinds of plants that grow in their community and assemble them in a computer file or print them and place them in a notebook. Students should label each photograph with key information, such as the name of the plant, its location, and the date of the photograph.

By using a mixture of formative and summative assessment types, including alternative assessments, you can get a more rounded picture of students' knowledge and skills in science.

Name _____ Date _____

Student book pages 002–005

Grade

A **TRUE/FALSE** Write if the sentence is true (T) or false (F). If the sentence is false, change the underlined word or phrase to make it true. *(20 points: 4 points each)*

Example: Scientists use **a descriptive design** to look for relationships. ___F___

a correlational design

1. Sometimes scientists describe **animals, plants, or rocks**. ____ _____

2. In **an experimental design**, a scientist forms a hypothesis. ____ _____

3. A scientist forms a hypothesis and then makes **conclusions**. ____ _____

4. Scientists **do experiments** to help answer questions. ____ _____

5. Scientists often **ask questions** about what they observe. ____ _____

B **MULTIPLE CHOICE** Choose the correct answer. *(20 points: 4 points each)*

Example: A scientist wonders, "Why doesn't the cell phone work?" The scientist is ___b___.
    a. predicting an outcome   **c.** observing nature
    b. asking a question   **d.** making a conclusion

6. A scientist observed what kind of food a bird eats. This is an example of ____.
    a. descriptive design   **c.** experimental design
    b. correlational design   **d.** a hypothesis

7. A scientist tests how long different materials keep ice frozen. The scientist is ____.
    a. asking a question   **c.** gathering data
    b. making a prediction   **d.** drawing a conclusion

8. A scientist looks for a relationship between dead fish and the water in a stream. This is an example of ____.
    a. descriptive design   **c.** experimental design
    b. correlational design   **d.** a conclusion

9. A scientist says, "My phone doesn't work. I think the battery is dead." The scientist is ____.
    a. gathering data   **c.** recording data
    b. testing a hypothesis   **d.** forming a hypothesis

10. A scientist studies a chart of birds' beaks and the foods the birds eat. The scientist is ____.
    a. testing a hypothesis   **c.** looking for relationships
    b. doing an experiment   **d.** asking a question

C **COMPLETION** Complete the sentences. *(20 points: 4 points each)*

Example: Scientists do experiments to answer their ____questions____.

11. A scientist notices dead fish in a stream. The scientist is making an _____.

12. A scientist thinks that something in a stream's water is killing fish. The scientist is

    forming a _____.

1

13. A scientist studies a bar graph and finds out which material keeps ice frozen longest. The scientist is making a _____.

14. A scientist wants to find the relationship between birds' beaks and the food they eat. The scientist is using _____ design.

15. A scientist tests different materials to see which keeps ice frozen longest. The scientist is using _____ design.

D SHORT ANSWERS Look at the chart. Answer the questions.
(20 points: 4 points each)

Sparrows have short, thick beaks for cracking seeds.	Woodpeckers have long, sharp beaks for drilling into wood to get insects, such as termites and ants.	Hummingbirds have long, thin beaks for reaching nectar in flowers.	Warblers have thin, pointed beaks for catching caterpillars and flying insects.

Example: A bird drills into wood. What kind of beak does it have?

_____long, sharp_____

16. A bird has a short, thick beak. What kind of food does it eat?

_____

17. What kind of beak do warblers have? _____

18. What kind of food do warblers eat? _____

19. What kind of bird has a very long, thin beak?

_____

20. What kind of food do hummingbirds eat?

_____

E WRITING You notice that different kinds of birds live in different kinds of homes. You wonder why. You think there is a relationship between a bird's size and the kind of home it lives in. What scientific method or methods will you use to find out? Write a paragraph explaining what you will do. *(20 points)*

_____

_____

_____

_____

_____

GATEWAY TO SCIENCE Assessment Book • Copyright © Thomson Heinle

Name _____ Date _____

Grade

A **TRUE/FALSE** Write if the sentence is true (T) or false (F). If the sentence is false, change the underlined word or phrase to make it true. *(20 points: 4 points each)*

Example: **An anemometer** measures temperature. __F__ ____a thermometer____

1. Scientists use **petri dishes** to watch for bad weather. ____ _____

2. A compound light microscope has an **eyepiece lens**. ____ _____

3. You can use a graduated cylinder to find the **temperature** of a rock. ____ _____

4. All science tools **are** in labs. ____ _____

5. Scientists can use **a balance** to measure things. ____ _____

B **MULTIPLE CHOICE** Choose the correct answer. *(20 points: 4 points each)*

Example: Scientists use test tubes __a__.
- **a.** to hold things
- **b.** to measure air temperature
- **c.** to take pictures of clouds
- **d.** to enlarge things

6. Scientists store data in ____.
- **a.** compound light microscopes
- **b.** geostationary satellites
- **c.** computers
- **d.** balances

7. When you magnify an object, you ____ it.
- **a.** record
- **b.** enlarge
- **c.** watch
- **d.** read

8. A ____ measures air temperature.
- **a.** microscope
- **b.** satellite
- **c.** beaker
- **d.** thermometer

9. A GOES weather satellite measures ____.
- **a.** Earth's orbit
- **b.** Earth's speed
- **c.** weather conditions
- **d.** cloud volume

10. A microscope helps us see ____ .
- **a.** small objects
- **b.** things that are far away
- **c.** things that are in a beaker
- **d.** things in a computer

C **COMPLETION** Complete the sentences. *(20 points: 4 points each)*

Example: You can use a graduated cylinder to measure the ____volume____ of a solid.

11. A scientist uses a _____ to look at planets.

12. Turn the coarse adjustment knob on a microscope to make an object clear. Then turn the _____ to make the object clearer.

13. A geostationary _____ stays above one place on Earth.

14. Before you magnify an object with a microscope, place the object on a glass

    _____.

15. A word that ends in *meter* can name a tool for _____.

D **SHORT ANSWERS** Look at the cylinders. Answer the questions.
(20 points: 4 points each)

Rock

Example: What do you call the curve in the surface of the water in a cylinder?

_____the meniscus_____

16. Which cylinder holds the bigger volume, the one on the right or the one on the left?

    _____

17. What caused one cylinder to hold a bigger volume?

    _____

18. You want to know the volume of the water in a cylinder. What part of the water do

    you look at? _____

19. Where is the largest number on a graduated cylinder?

    _____

20. What do you subtract from the volume of the water and the rock to find the volume

    of the rock? _____

E **WRITING** You want to describe the weather today. What will you measure? What
tools will you use? Write a paragraph. *(20 points)*

_____

_____

_____

_____

_____

GATEWAY TO SCIENCE Assessment Book • Copyright © Thomson Heinle

Name _____ Date _____

📖 Student book pages 010–013

Grade ☐

A **TRUE/FALSE** Write if the sentence is true (T) or false (F). If the sentence is false, change the <u>underlined</u> word or phrase to make it true. *(20 points: 4 points each)*

Example: The base unit for measuring volume is the **gram**. __F__ _____liter_____

1. Most people in the United States use the <u>**Fahrenheit scale**</u> to measure temperature. ____ _____

2. A kilometer is one **hundred** meters. ____ _____

3. **Volume** tells how much matter is in an object. ____ _____.

4. A **unit** is an amount that never changes. ____ _____

5. Before the metric system, people in different countries used **different** measurement systems. ____ _____

B **MULTIPLE CHOICE** Choose the correct answer. *(20 points: 4 points each)*

Example: The base unit for measuring mass is the __C__.
   a. liter   b. meter   c. gram   d. degree

6. A centimeter is ____.
   a. one hundred (100) meters   c. one thousandth (0.001) of a meter
   b. one thousand (1000) meters   d. one hundredth (0.01) of a meter

7. Which measurement is not part of the metric system? ____
   a. gram   b. mile   c. meter   d. centiliter

8. Scientists measure temperature using the ____.
   a. metric system   c. Celsius scale
   b. Fahrenheit scale   d. United States system

9. Distances between towns are measured in ____.
   a. centimeters   b. millimeters   c. kilograms   d. kilometers

10. Centiliters and milliliters are measures of ____.
   a. mass   b. volume   c. temperature   d. length

C **COMPLETION** Complete the sentences. *(20 points: 4 points each)*

Example: The letters mL stand for the word _____milliliter_____.

11. Scientists measure _____ in centimeters.

12. The liter is the base unit for measuring _____.

13. The number 0.003 in words is _____.

14. Scientists use a _____ to measure how hot something is.

15. A millimeter is _____ of a meter.

D **SHORT ANSWERS** Look at the thermometers. Answer the questions.
(20 points: 4 points each)

Temperature Scales

°F		°C
230		110
Water→212		100←Water
boils 194		90 boils
176		80
158		70
140		60
122		50
104		40
86		30
68		20
50		10
Water→32		0←Water
freezes		freezes

Fahrenheit Celsius

Example: At what temperature does water freeze on the Celsius scale?

_____ *zero degrees* _____

16. At what temperature does water boil on the Fahrenheit scale?

_____

17. At what temperature does water boil on the Celsius scale?

_____

18. At what temperature does water freeze on the Fahrenheit scale?

_____

19. Which temperature scale has more degrees between the freezing point and the

boiling point of water? _____

20. The temperature today is 68 degrees Fahrenheit. What is the temperature in degrees

Celsius? _____

E **WRITING** When you see the doctor, a nurse will measure your body. The doctor
wants to know how tall you are, your mass, and your temperature. Write a paragraph about the units the nurse will use. *(20 points)*

_____

_____

_____

_____

_____

GATEWAY TO SCIENCE Assessment Book • Copyright © Thomson Heinle

Name _____ Date _____

Grade

A **TRUE/FALSE** Write if the sentence is true (T) or false (F). If the sentence is false, change the underlined word or phrase to make it true. *(20 points: 4 points each)*

Example: A **bar graph** connects points with a line. __F__ ____line graph____

1. A pie chart shows data in **percentages.** ____ _____

2. A **Venn diagram** shows how things are the same and different. ____ _____

3. The parts of a **table** add up to 100 percent. ____ _____

4. A Venn diagram contains **two squares**. ____ _____

5. A **line graph** can show changes over time. ____ _____

B **MULTIPLE CHOICE** Choose the correct answer. *(20 points: 4 points each)*

Example: A __d__ of a table tells the data that is being compared.
   a. row   **b.** column   **c.** line   **d.** title

6. A pie chart shows a ____ divided into parts.
   a. map key   **b.** bar   **c.** circle   **d.** map

7. A ____ can show steps in a process.
   a. bar graph   **b.** Venn diagram   **c.** weather map   **d.** flowchart

8. The columns in a table go ____.
   a. across   **c.** from point to point
   b. up and down   **d.** to 100 percent

9. Tables and graphs can help scientists ____ data.
   a. change   **b.** divide   **c.** compare   **d.** make

10. A ____ can show the surface of Earth.
   a. map   **b.** pie chart   **c.** line graph   **d.** Venn diagram

C **COMPLETION** Complete the sentences. *(20 points: 4 points each)*

Example: The _____data_____ in a table can be in numbers or in words.

11. A line graph shows the _____ between numbers.

12. The rows in a table go _____.

13. A _____ graph connects points with a line.

14. The parts of a pie chart are called _____.

15. The parts of a pie chart add up to _____ percent.

Name _____   Date _____

D **SHORT ANSWERS** Look at the bar graph. Answer the questions.
(20 points: 4 points each)

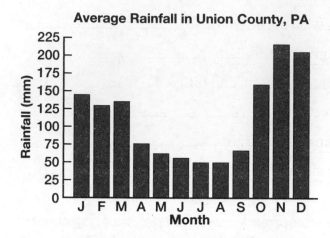

The months in a year are January, February, March, April, May, June, July, August, September, October, November, December.

Example: What month had the most rainfall? ___November___

16. Which two months had the least rainfall?
    _____

17. Which month had more rainfall, January or February?
    _____

18. Which month had less rainfall, March or October?
    _____

19. Which months had more rainfall than October?
    _____

20. Did April have more or less rainfall than May?
    _____

E **WRITING** You want to show how temperature changes in your town during May. What data will you collect to find out? How will you organize that data? What kind of graph will you make? Describe what you will do. Write a paragraph. *(20 points)*

_____

_____

_____

_____

_____

GATEWAY TO SCIENCE Assessment Book • Copyright © Thomson Heinle

Name _____   Date _____

☐ Student book pages 018–021   Grade ☐

A   TRUE/FALSE   Write if the sentence is true (T) or false (F). If the sentence is false, change the underlined word or phrase to make it true. *(20 points: 4 points each)*

Example: If you hurt yourself in the lab, tell your **friends.** __F__   __teacher__

1. If you cut yourself, you will need a **fire extinguisher.** ____ _____

2. A **hot plate** can burn you. ____ _____

3. You wear goggles in the lab to protect your **ears.** ____ _____

4. If you touch a chemical, **ring the fire alarm** right away. ____ _____

5. A lab apron can help protect your **clothes.** ____ _____

B   MULTIPLE CHOICE   Choose the correct answer. *(20 points: 4 points each)*

Example: Wear heatproof gloves when working with ___b___.
   a. soap and water   **b.** a hot plate   **c.** chemicals   **d.** a first aid kit

6. Some living things scientists study are ____.
   a. chemicals and plants   **c.** flames and birds
   b. animals and water   **d.** animals and plants

7. If you burn your hand, ____.
   a. wash it in cold water   **c.** put on a heatproof glove
   b. wash it in hot water   **d.** ring the fire alarm

8. When you heat a liquid in a beaker, ____.
   a. ring the fire alarm   **c.** use a ring stand and a wire gauze
   b. use soap and water   **d.** use a first aid kit

9. When you study living things, ____.
   a. wear heatproof gloves   **c.** always take them to a lab
   b. do not harm them   **d.** wrap them in a fire blanket

10. When you prevent accidents, you ____.
   a. remove them   **b.** harm them   **c.** avoid them   **d.** protect them

C   COMPLETION   Complete the sentences. *(20 points: 4 points each)*

Example: You can use a _____first aid kit_____ to treat burns.

11. It is important not to harm _____ things.

12. If you spill chemicals on your hand, wash it with _____ and water.

13. Don't bring living things to class. Take _____ of them.

14. You should always follow _____.

15. Read warnings and _____ to help prevent accidents.

D **SHORT ANSWERS** Look at the safety signs. Answer the questions.
(20 points: 4 points each)

Safety Signs and Warnings

Example: What does the electric shock sign tell you?

_____ *be careful with electricity* _____

16. What can cause you to cut yourself? _____

17. What sign shows a beaker? _____

18. Why shouldn't you touch something that has an Electric Shock sign?

_____

19. What sign shows a hand? _____

20. What should you wear when you see the Corrosive Chemicals sign?

_____

E **WRITING** You are doing an experiment, and you burn your finger. What will you do? Write a paragraph. *(20 points)*

_____

_____

_____

_____

_____

GATEWAY TO SCIENCE Assessment Book • Copyright © Thomson Heinle

Name _____ Date _____

📖 Student book pages 002–021

Grade

A **TRUE/FALSE** Write if the sentence is true (T) or false (F). If the sentence is false, change the underlined word or phrase to make it true. *(20 points: 2 points each)*

Example: Scientists use **a descriptive design** to look for relationships.

   F   *a correlational design*

1. Scientists sometimes describe **animals, plants, or rocks**. ____ _____

2. In **an experimental design**, a scientist forms a hypothesis. ____ _____

3. Scientists use **test tubes** to watch for bad weather. ____ _____

4. A compound light microscope has **objective lenses**. ____ _____

5. Most people in the United States use the **Fahrenheit scale** to measure temperature.

   ____ _____

6. A kilometer is one **hundred** meters. ____ _____

7. A pie chart shows data in **percentages**. ____ _____

8. A **Venn diagram** shows how things are alike and different. ____ _____

9. If you cut yourself, you will need a **fire alarm**. ____ _____

10. **Hot glass** can burn you. ____ _____

B **MULTIPLE CHOICE** Choose the correct answer. *(40 points: 2 points each)*

Example: A scientist wonders, "Why doesn't the cell phone work?" The scientist is __b__.
  a. predicting a hypothesis   **c.** observing nature
  b. asking a question   **d.** making a conclusion

11. A scientist observes what kind of food a bird eats. This is an example of ____.
  a. descriptive design   **c.** experimental design
  b. correlational design   **d.** a conclusion

12. A scientist tests how long different materials keep ice frozen. The scientist is ____.
  a. asking a question   **c.** gathering data
  b. forming a hypothesis   **d.** drawing a conclusion

13. A scientist looks for a relationship between dead fish and the water in a stream. This is an example of a ____.
  a. descriptive design   **c.** experimental design
  b. correlational design   **d.** conclusion

B MULTIPLE CHOICE, continued

14. A scientist says, "My cell phone doesn't work. I think the battery is dead." The scientist is _____.

    a. gathering data
    b. testing a hypothesis

    c. recording data
    d. forming a hypothesis

15. Scientists store data in _____.

    a. compound light microscopes
    b. graduated cylinders

    c. computers
    d. balances

16. When you magnify an object, you _____ it.

    a. analyze   **b.** enlarge   **c.** watch   **d.** read

17. A _____ measures air temperature.

    a. balance   **b.** satellite   **c.** beaker   **d.** thermometer

18. A GOES weather satellite measures _____.

    a. Earth's volume
    b. Earth's speed

    c. weather conditions
    d. cloud volume

19. A centimeter is _____.

    a. one hundred (100) meters
    b. one thousand (1000) meters

    c. one thousandth (0.001) of a meter
    d. one hundredth (0.01) of a meter

20. Which measurement is not part of the metric system? _____

    a. gram   **b.** mile   **c.** meter   **d.** centimeter

21. Scientists measure temperature using the _____.

    a. metric system
    b. Fahrenheit scale

    c. Celsius scale
    d. English system

22. Distances between towns are measured in _____.

    a. centimeters   **b.** decimeters   **c.** kilograms   **d.** kilometers

23. A pie chart shows a _____ divided into parts.

    a. Venn diagram   **b.** bar   **c.** circle   **d.** map

24. A _____ can show steps in a process.

    a. line graph   **b.** Venn diagram   **c.** weather map   **d.** flowchart

25. The columns in a table go _____.

    a. across
    b. up and down

    c. from tallest to shortest
    d. to 100 percent

26. Tables and graphs can help scientists _____ data.

    a. change   **b.** divide   **c.** compare   **d.** predict

GATEWAY TO SCIENCE Assessment Book • Copyright © Thomson Heinle

B **MULTIPLE CHOICE, continued**

27. Some living things scientists study are ____.
   a. chemicals and plants   **c.** patterns and birds
   b. animals and water   **d.** animals and plants

28. If you burn your hand, ____.
   a. wash it in cold water   **c.** use a ring stand
   b. wash it in hot water   **d.** ring the fire alarm

29. When you heat a liquid in a beaker, ____.
   a. ring the fire alarm   **c.** use a ring stand and a wire gauze
   b. use soap and water   **d.** use a safety sign

30. When you study living things, ____.
   a. wear heatproof gloves   **c.** always take them to a lab
   b. do not harm them   **d.** wear an apron

C **COMPLETION** Complete the sentences. *(20 points: 2 points each)*

Example: A bacteria cell has a cell membrane and _____cytoplasm_____.

31. A scientist notices dead fish in a stream. The scientist is making an _____.

32. A scientist thinks that something in a stream's water is killing fish. The scientist is

   forming a _____.

33. A scientist uses a _____ to look at moons.

34. Turn the coarse adjustment knob on a microscope to make an object clear. Then turn

   the _____ to make the object clearer.

35. Scientists measure _____ in centimeters.

36. The liter is the base unit for measuring _____.

37. A line graph shows the _____ between numbers.

38. The rows in a table go _____.

39. It is important not to harm _____ things.

40. If you spill chemicals on your hand, wash it with _____ and water.

Name _____ Date _____

D **SHORT ANSWERS** Look at the chart. Answer the questions.
(20 points: 2 points each)

Sparrows have short, thick beaks for cracking seeds.	Woodpeckers have long, sharp beaks for drilling into wood to get insects, such as termites and ants.	Hummingbirds have long, thin beaks for reaching nectar in flowers.	Warblers have thin, pointed beaks for catching caterpillars and flying insects.

Example: A bird drills into wood to get insects to eat. What kind of beak does it have?

_____long, sharp_____

41. A bird has a short, thick beak. What kind of food does it eat?

_____

42. What kind of beak do warblers have?

_____

43. What kind of food do warblers eat?

_____

44. What kind of bird has a very long, thin beak?

_____

45. What kind of food do hummingbirds eat?

_____

GATEWAY TO SCIENCE Assessment Book • Copyright © Thomson Heinle

Name _____ Date _____

D SHORT ANSWERS, continued

Look at the cylinders. Answer the questions.

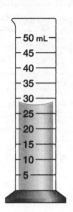

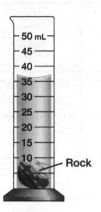

46. Which cylinder holds the bigger volume, the one on the right or the one on the left?

    _____

47. What caused one cylinder to hold a bigger volume?

    _____

48. You want to know the volume of the water in a cylinder. What part of the water do

    you look at? _____

49. Where is the largest number on a graduated cylinder?

    _____

50. What do you subtract from the volume of the water and the rock to find the volume

    of the rock? _____

E **WRITING ASSESSMENT** Write paragraphs. *(100 points: 25 points each)*

51. A scientist sees some dead trees. The ground around the trees is covered with oil. What question might the scientist ask? What three steps can the scientist take to find out why the trees died?

_____

_____

_____

_____

_____

52. You have two small objects. Which has the bigger volume? Describe how you would find out.

_____

_____

_____

_____

_____

53. Why was the invention of the metric system important in history?

_____

_____

_____

_____

_____

54. You want to tell how cats and dogs are similar and different. What kind of chart or graph will you use? Explain the parts of the chart or graph you will use. What kind of information will you write in each part?

_____

_____

_____

_____

_____

GATEWAY TO SCIENCE Assessment Book • Copyright © Thomson Heinle

Name _____   Date _____

Grade

A **TRUE/FALSE** Write if the sentence is true (T) or false (F). If the sentence is false, change the <u>underlined</u> word or phrase to make it true. *(20 points: 4 points each)*

Example: <u>Ribosomes</u> store water, food, and waste. __F__ _____vacuoles_____

1. <u>Lysosomes</u> break down material. ____ _____

2. Your body is made up of many <u>Golgi</u>. ____ _____

3. All cells have a <u>cell wall</u>. ____ _____

4. The <u>cell</u> is the basic unit of living things. ____ _____

5. The word part *plastikos* means <u>"to form or mold."</u> ____ _____

B **MULTIPLE CHOICE** Choose the correct answer. *(20 points: 4 points each)*

Example: The __b__ packages proteins.

  a. cytoplasm   **b.** Golgi complex   **c.** cell wall   **d.** nucleus

6. Robert Hooke saw ____.

  a. the walls of a cell   **c.** the organelles of a cell
  b. the nucleus of a cell   **d.** a Golgi

7. The ____ is a jelly-like material.

  a. cytoplasm   **b.** mitochondria   **c.** chloroplast   **d.** nucleus

8. The nucleus of a cell is in charge of cell activities. This means it ____.

  a. digests material in the cell   **c.** controls activities in the cell
  b. holds up cell walls   **d.** processes proteins in the cell

9. Mitochondria ____.

  a. control cell activities   **c.** package proteins
  b. make energy   **d.** hold up the plant

10. Ribosomes ____.

  a. control activity in the cell   **c.** digest material
  b. build proteins   **d.** control what goes in and out of the cell

C **COMPLETION** Complete the sentences. *(20 points: 4 points each)*

Example: All living things are made of one or more _____cell_____.

11. The cell parts that perform life activities are called _____.

12. The organelles that make food in green plants are _____.

13. A _____ cell does not have a nucleus.

14. Robert Hooke looked at cells under a _____.

15. A stiff _____ holds up a plant.

Name _____ Date _____

D SHORT ANSWERS Look at the table. Answer the questions.
(20 points: 4 points each)

	Animal Cell	Plant Cell	Bacteria Cell
cell membrane	√	√	√
cytoplasm	√	√	√
chloroplasts		√	
cell wall		√	

Example: What kind of cell has the most kinds of organelles?

_____ a plant cell _____

16. What kind of cell has the least kinds of organelles?

_____

17. How many kinds of organelles does a bacteria cell have?

_____

18. How many more kinds of organelles does a plant cell have than an animal cell?

_____

19. How many columns does the table have?

_____

20. What kind of cell has chloroplasts? _____

E WRITING Tell about Robert Hooke's work. What did he study? What tool did he use? What did he see? What did he invent? Write a paragraph about him.
(20 points)

_____

_____

_____

_____

_____

Name _____ Date _____

Grade ☐

A **TRUE/FALSE** Write if the sentence is true (T) or false (F). If the sentence is false, change the <u>underlined</u> word or phrase to make it true. *(20 points: 4 points each)*

Example: Most living things are made of **many cells**. __F__ ____one cell____

1. All one-celled living things need **sunshine**. ____ _____

2. The green chemical in **flagella** helps some cells make their own food. ____

_____

3. Protozoans get food by eating other **cells**. ____ _____

4. Some bacteria live in places with no **air**. ____ _____

5. **Algae** use sunlight to make their own food. ____ _____

B **MULTIPLE CHOICE** Choose the correct answer. *(20 points: 4 points each)*

Example: People use __c__ to make bread.
   a. cilia   **b.** pseudopods   **c.** yeast   **d.** bacteria

6. A trichonympha is a kind of ____.
   a. protozoan   **b.** yeast   **c.** pseudopod   **d.** tube worm

7. Some trypanosomes get energy by eating ____.
   a. chloroplasts   **b.** sunlight   **c.** water   **d.** blood cells

8. Tube worms can live in very ____ ocean water.
   a. cold   **b.** hot   **c.** salty   **d.** shallow

9. What does the word part *pod* in *pseudopod* mean?
   a. spiral   **b.** vacuole   **c.** foot   **d.** rod

10. All one-celled organisms live in ____ places.
   a. dry   **b.** salty   **c.** hot   **d.** wet

C **COMPLETION** Complete the sentences. *(20 points: 4 points each)*

Example: There are rod, spiral, and ____round____ bacteria.

11. Yeast is a one-celled kind of _____.

12. Algae are grouped as red, gold, or _____.

13. Some bacteria live inside giant _____ worms.

14. A single bacteria cell is called a _____.

15. Some bacteria can live in water that is very hot or water that is very

_____.

Name _____ Date _____

D SHORT ANSWERS Look at the pictures. Answer the questions.
(20 points: 4 points each)

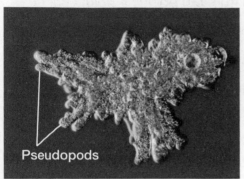

Pseudopods

An amoeba uses pseudopods to move and get food.

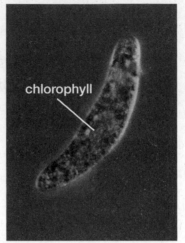

chlorophyll

A euglena can find food or make its own food with sunlight and chlorophyll.

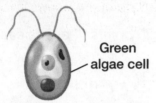

Green algae cell

Green algae make their own food. They get energy from sunlight.

Example: What does an amoeba use its pseudopods for?
_____ *to move and to get food* _____

16. What does a euglena use to make food with sunlight?

_____

17. Which cell must always find its food?

_____

18. When does a euglena find its food? _____

19. Which organisms are most like plants?

_____

20. Which organism is most like an animal in the way it gets food?

_____

E WRITING What are some extreme places where bacteria can live? Describe those places. Write a paragraph. (20 points)

_____

_____

_____

_____

GATEWAY TO SCIENCE Assessment Book • Copyright © Thomson Heinle

Name _____ Date _____

📖 Student book pages 030–033

Grade ☐

A **TRUE/FALSE** Write if the sentence is true (T) or false (F). If the sentence is false, change the underlined word or phrase to make it true. *(20 points: 4 points each)*

Example: Details are **large**, important pieces of information __F__ __small__

1. The lungs and the trachea are part of the **respiratory system**. ____ _____

2. A multicellular animal contains **one kind** of cell. ____ _____

3. In cell division, a cell copies its **nucleus**. ____ _____

4. **Tissues** are groups of the same kinds of cells. ____ _____

5. **Red blood cells** hunt and eat bacteria. ____ _____

B **MULTIPLE CHOICE** Choose the correct answer. *(20 points: 4 points each)*

Example: Leaves, stems, and flowers are part of a plant's __b__.
    a. respiratory system   **c.** roots
    b. shoot system   **d.** lung tissue

6. In cell division, an organism ____.
    a. collects gases   **c.** makes new cells
    b. hunts and eats bacteria   **d.** helps move bones

7. Chromosomes contain information on ____.
    a. how bone cells help the body   **c.** how the heart beats
    b. how air moves in the lungs   **d.** how to make new cells

8. In cell division, two sets of chromosomes ____.
    a. line up and move away from each other   **c.** form new organs
    b. hunt and eat each other   **d.** make a tissue

9. All the cells in a tissue ____.
    a. do the same kind of work   **c.** do different jobs
    b. help protect the body   **d.** move away from each other

10. Two or more tissues working together form ____.
    a. a chromosome   **b.** a nucleus   **c.** an organ   **d.** a white blood cell

C **COMPLETION** Complete the sentences. *(20 points: 4 points each)*

Example: The word part *multi* in *multicellular* means __much or many__.

11. Tubes in plants carry things between the roots and the _____.

12. White blood cells protect the body against _____.

13. After cell division, each new cell has one set of _____.

14. _____ tissue helps people breathe.

15. _____ cells protect the body from injury and germs.

D **SHORT ANSWERS** Look at the chart. Answer the questions.
(20 points: 4 points each)

Kind of Cell		What does this kind of cell do?
skeletal muscle cell		Skeletal muscle cells are connected to bones. This kind of cell forms tissues that help move bones.
smooth muscle cell		This kind of cell forms organs that help move food in the digestive system. It also lines blood vessels and the airways of the lungs.
bone cell		The skeleton is made up of bone cells. This kind of cell provides support and allows the body to move.
red blood cell		This kind of cell carries gases to other cells in the body.

Example: What are skeletal muscle cells connected to?

_____ *bones* _____

16. What kind of cells is the skeleton made of?

_____

17. What kind of cells line blood vessels?

_____

18. What kind of cells provide support for the body?

_____

19. What do smooth muscle cells do in the digestive system?

_____

20. What kind of cells carry gases to other cells in the body?

_____

E **WRITING** What is a tissue? What is it made of? Give examples of tissues. Write a paragraph about tissues. *(20 points)*

_____

_____

_____

_____

_____

Grade

A **TRUE/FALSE** Write if the sentence is true (T) or false (F). If the sentence is false, change the underlined word to make it true. *(20 points: 4 points each)*

Example: **Clay** is made from dead plants. ___F___ _____humus_____

1. **Roots** hold a plant in the ground. ____ _____

2. A fruit may contain one or more **spores**. ____ _____

3. A cone protects the **roots** of a plant. ____ _____

4. Plants store food in their **stems**. ____ _____

5. **Soil** contains air and water. ____ _____

B **MULTIPLE CHOICE** Choose the correct answer. *(20 points: 4 points each)*

Example: Nitrogen is ___c___.
  a. a spore   **b.** a kind of soil   **c.** a nutrient   **d.** a moss

6. Leaves use energy from ____ to make food.
  a. roots   **b.** sunlight   **c.** spores   **d.** flowers

7. What plant part carries water from the roots to the leaves? ____
  a. the stem   **b.** the flowers   **c.** the cones   **d.** the seeds

8. Water runs quickly through ____.
  a. loam   **b.** clay   **c.** sand   **d.** nitrogen

9. The leaves of a Venus flytrap can digest ____.
  a. humus   **b.** insects   **c.** spores   **d.** minerals

10. Spores are cells that ____.
  a. make new plants   **c.** hold a plant in place
  b. make food from sunlight   **d.** make fruits

C **COMPLETION** Complete the sentences. *(20 points: 4 points each)*

Example: Plant roots take in water and minerals from _____soil_____.

11. Humus is rich in the _____ plants need to grow.

12. Water and air cannot flow easily through _____ soil.

13. The Venus flytrap grows where the soil has no _____.

14. Soil that has humus, sand, clay, air, and water is called _____.

15. The small pieces of rock in soil provide _____ to plants.

D **SHORT ANSWERS** Look at the bar graph. Answer the questions.
(20 points: 4 points each)

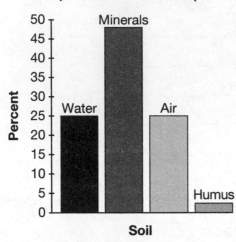

Composition of a Soil Sample

Example: Does this soil sample have more minerals or more water?

_____*more minerals*_____

16. What material makes up the smallest part of this soil sample?

_____

17. About what percent of the soil sample is humus?

_____

18. What two things does the sample have the same amount of?

_____

19. About what percent of the soil sample is minerals?

_____

20. What percent of the soil sample is air and water?

_____

E **WRITING** How is a Venus flytrap like other plants? What nutrient does a flytrap need? Where does it get that? How is a Venus flytrap different from other plants? Write a paragraph. *(20 points)*

_____

_____

_____

_____

_____

Name _____ Date _____

📖 Student book pages 038–041

Grade

A TRUE/FALSE Write if the sentence is true (T) or false (F). If the sentence is false, change the underlined word or phrase to make it true. *(20 points: 4 points each)*

Example: Vascular plants have tubes in their **seeds**. __F__ ____stems____

1. Conifers are **seedless** plants that produce cones. ____ _____

2. The anther of a flowering plant makes **spores**. ____ _____

3. Ferns and **mosses** are seedless plants. ____ _____

4. Flowering plants reproduce by making **spores**. ____ _____

5. A fern plant does **not have** roots. ____ _____

B MULTIPLE CHOICE Choose the correct answer. *(20 points: 4 points each)*

Example: When a seed is ready to germinate, it absorbs __c__.
   a. pollen   **b.** spores   **c.** water   **d.** wind

6. A flowering plant's ___ contains eggs.
   a. anther   **b.** stem   **c.** petals   **d.** pistil

7. The ___ of a flowering plant move water and minerals from the roots to the leaves.
   a. seeds   **b.** spore   **c.** xylem   **d.** phloem

8. Conifers are ___ plants.
   a. seedless   **b.** nonvascular   **c.** vascular   **d.** rootless

9. Nonvascular plants often live where it is dark and ___.
   a. dry   **b.** sunny   **c.** sticky   **d.** wet

10. The new plant that grows from a seed is ___.
   a. a seedling   **b.** a moss   **c.** a spore   **d.** an anther

C COMPLETION Complete the sentences. *(20 points: 4 points each)*

Example: Xylem and phloem are ____tubes____ in a vascular plant's stem.

11. A flower contains stamens and one or more _____.

12. When pollen lands on a pistil, _____ happens.

13. Nonvascular plants do not have tubes or _____.

14. When a seed _____, it starts growing.

15. Ferns reproduce by making _____.

D **SHORT ANSWERS** Look at the flow chart. Answer the questions.
(20 points: 4 points each)

Reproduction of Flowering Plants

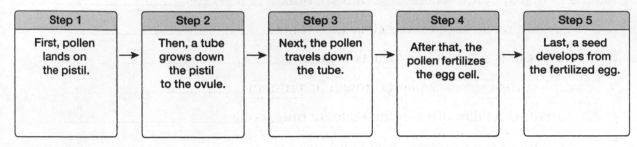

Step 1	Step 2	Step 3	Step 4	Step 5
First, pollen lands on the pistil.	Then, a tube grows down the pistil to the ovule.	Next, the pollen travels down the tube.	After that, the pollen fertilizes the egg cell.	Last, a seed develops from the fertilized egg.

Example: In what step does pollination happen?

_____Step 1_____

16. How does pollen reach the egg cell?

_____

17. Where is the ovule? _____

18. Does the pollen travel down the tube before or after it lands on the pistil?

_____

19. Is the egg fertilized before or after the pollen travels down the tube?

_____

20. What happens after the pollen fertilizes the egg cell?

_____

E **WRITING** A seed falls to the ground and begins to germinate. Describe how it happens. What happens after that? Write a paragraph. *(20 points)*

_____

_____

_____

_____

_____

GATEWAY TO SCIENCE Assessment Book • Copyright © Thomson Heinle

Name _____ Date _____

Grade

A **TRUE/FALSE** Write if the sentence is true (T) or false (F). If the sentence is false, change the <u>underlined</u> word to make it true. *(20 points: 4 points each)*

Example: Photosynthesis takes place in the **stomata** of leaf cells. __F__ __chloroplasts__

1. Guard cells in leaves open and close the **stomata**. ____ _____

2. Photosynthesis produces **oxygen** as waste. ____ _____

3. Water in the soil enters a plant through its **phloem**. ____ _____

4. Photosynthesis takes place in the **stomata** of leaf cells. ____ _____

5. Many kinds of organisms eat **plants** for food. ____ _____

B **MULTIPLE CHOICE** Choose the correct answer. *(20 points: 4 points each)*

Example: The outer bark of a tree is made of dead __a__.
    a. phloem cells   **b.** glucose   **c.** chloroplasts   **d.** guard cells

6. The word part *photo* in *photosynthesis* means ____.
    a. to place   **b.** oxygen   **c.** together   **d.** light

7. A group of cells that do the same job are ____.
    a. an organism   **b.** glucose   **c.** a trunk   **d.** a tissue

8. Carbon dioxide and oxygen enter and leave a plant through ____.
    a. xylem   **b.** phloem   **c.** stomata   **d.** roots

9. The annual rings in a tree trunk are made of old ____.
    a. xylem and phloem tissue   **c.** guard cells
    b. roots   **d.** chloroplasts

10. What plant part uses the energy in sunlight to make food? ____
    a. stems   **b.** leaves   **c.** phloem   **d.** stomata

C **COMPLETION** Complete the sentences. *(20 points: 4 points each)*

Example: You can tell the age of a tree by counting its ____annual rings____.

11. Wide rings in a tree trunk are a sign of a year with a lot of _____.

12. Plants are organisms that can make their own _____.

13. The glucose that plants make is a kind of _____.

14. Stomata are found on the underside of plant _____.

15. Phloem tissue carries food to a plant's _____, stems, and leaves.

D **SHORT ANSWERS** Look at the diagrams. Answer the questions.
(20 points: 4 points each)

Diagram A

carbon dioxide + water $\xrightarrow[\text{chlorophyll}]{\text{sunlight}}$ sugar + oxygen

Diagram B

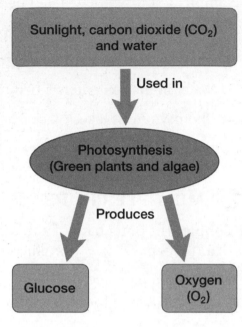

Sunlight, carbon dioxide (CO_2) and water

Used in

Photosynthesis (Green plants and algae)

Produces

Glucose

Oxygen (O_2)

Example: In which organisms does photosynthesis happen?

_____*green plants and algae*_____

16. What is CO_2? _____

17. What is O_2? _____

18. What does a plant use in photosynthesis?

_____

19. Diagram A has a step called "sunlight/chlorophyll." What is that step called in

Diagram B? _____

20. What kind of sugar is produced? _____

E **WRITING** You cut down a tree in your yard. How can you tell the tree's age? Was it dry or rainy last year? How can you tell? Write a paragraph. *(20 points)*

_____

_____

_____

_____

_____

GATEWAY TO SCIENCE Assessment Book • Copyright © Thomson Heinle

Name _____ Date _____

Student book pages 046–049

Grade

A) TRUE/FALSE Write if the sentence is true (T) or false (F). If the sentence is false, change the <u>underlined</u> word to make it true. *(20 points: 4 points each)*

Example: An octopus eats **mice** for food. _F_ _____shellfish_____

1. When animals **reproduce**, they make new baby animals. ____ _____

2. Animals need **water** for energy. ____ _____

3. **Most** animals have backbones. ____ _____

4. All animals are made of many **cells**. ____ _____

5. An octopus has **excellent** eyesight. ____ _____

B) MULTIPLE CHOICE Choose the correct answer. *(20 points: 4 points each)*

Example: Water helps keep animals __c__.
  a. dry   **b.** cool   **c.** soft   **d.** tough

6. Animals such as ____ have tough outer shells.
  a. opossums   **b.** owls   **c.** worms   **d.** insects

7. Most animals have ____.
  a. senses   **b.** outer shells   **c.** suckers   **d.** backbones

8. An octopus uses its arms ____.
  a. to see   **b.** to catch food **c.** to get water   **d.** to get oxygen

9. A backbone supports an animal's ____.
  a. shelter   **b.** brain   **c.** senses   **d.** body

10. An octopus uses its ____ to move quickly.
  a. suckers   **b.** funnel   **c.** ink   **d.** arms

C) COMPLETION Complete the sentences. *(20 points: 4 points each)*

Example: A crab has a tough outer _____shell_____.

11. An animal's _____ help it find food, water, and shelter.

12. On octopus uses its _____ to walk on the ocean floor.

13. Birds and raccoons can find shelter in _____.

14. The word part *octo* in *octopus* means _____.

15. An owl eats mice. Mice are _____ for owls.

Name _____ Date _____

D SHORT ANSWERS Look at the pie chart. Answer the questions.
(20 points: 4 points each)

Kinds of Animals

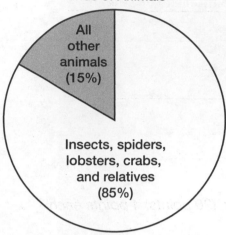

Example: What does this pie chart show?

_____kinds of animals_____

16. What is the total percentage of animals shown in the pie chart?

_____

17. Are raccoons members of the larger group or of the smaller group of animals?

_____

18. Are owls and mice members of the same group or different groups?

_____

19. Are frogs part of the larger or the smaller group of animals?

_____

20. Are fish and crabs members of the same group or different groups?

_____

E WRITING Animals have a number of common features. An owl is an animal.
Explain how you know that. Give examples of the common features that owls have.
Write a paragraph. *(20 points)*

_____

_____

_____

_____

_____

GATEWAY TO SCIENCE Assessment Book • Copyright © Thomson Heinle

Name _____ Date _____

Grade

Student book pages 050–053

A **TRUE/FALSE** Write if the sentence is true (T) or false (F). If the sentence is false, change the <u>underlined</u> word or phrase to make it true. *(20 points: 4 points each)*

Example: Invertebrates are most common **on land**. _F_ ____in the sea____

1. All arachnids have <u>six</u> legs. ____ _____

2. Sponges get food from <u>water</u>. ____ _____

3. <u>Invertebrates</u> are animals that have backbones. ____ _____

4. Scorpions are <u>arachnids</u>. ____ _____

5. A butterfly starts as <u>an egg</u>. ____ _____

B **MULTIPLE CHOICE** Choose the correct answer. *(20 points: 4 points each)*

Example: A butterfly egg hatches into a _b_.
   a. scorpion   **b.** caterpillar   **c.** sponge   **d.** chrysalis

6. A spider is a type of ____.
   a. sponge   **b.** jellyfish   **c.** beetle   **d.** arachnid

7. A beetle has ____.
   a. a hard body case   **b.** a backbone   **c.** eight legs   **d.** a chrysalis

8. Heartworms and tapeworms live inside ____.
   a. the sea   **b.** buildings   **c.** shells   **d.** other animals

9. A sponge pulls in water through its ____.
   a. body case   **b.** shell   **c.** pores   **d.** sections

10. Spiders and other arachnids ____.
   a. cannot move   **c.** develop through metamorphosis
   b. lay eggs   **d.** become butterflies

C **COMPLETION** Complete the sentences. *(20 points: 4 points each)*

Example: In the process of ____molting____, an arachnid loses its case and grows a new one.

11. A sponge's body is _____ at the bottom.

12. An arachnid's _____ is divided into two sections.

13. _____ leaves a sponge's body through the top opening.

14. A caterpillar changes into a butterfly inside a _____.

15. A sponge's body is a hollow _____.

D **SHORT ANSWERS** Look at the diagram. Answer the questions.
(20 points: 4 points each)

Metamorphosis

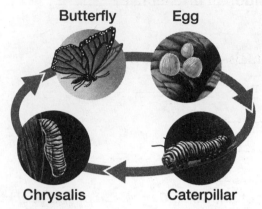

Butterfly   Egg

Chrysalis   Caterpillar

Example: Where do the eggs come from?

_____from the butterfly_____

16. What happens to an egg in the diagram?

_____

17. What happens to the caterpillar after it eats and grows?

_____

18. What happens inside the chrysalis? _____

19. How many stages are there in a butterfly's life?

_____

20. What is the name of the process shown in the diagram?

_____

E **WRITING** Imagine that you are a young sponge. What happens when you become an adult? How does your life change? What will you do? How will you get food? Write a paragraph. *(20 points)*

_____

_____

_____

_____

_____

Name _____ Date _____

Student book pages 054–057

Grade ☐

A TRUE/FALSE Write if the sentence is true (T) or false (F). If the sentence is false, change the <u>underlined</u> word to make it true. *(20 points: 4 points each)*

Example: Seahorses belong to the **amphibian** group of invertebrates. __F__

_____fish_____

1. The body temperature of **cold-blooded** animals always stays the same. ____

_____.

2. Humans are **mammals**. ____ _____

3. A kangaroo is a **reptile**. ____ _____

4. Fish are **cold-blooded**. ____ _____

5. Fish have **backbones**. ____ _____

B MULTIPLE CHOICE Choose the correct answer. *(20 points: 4 points each)*

Example: A penguin is a kind of __c__.
   a. fish   **b.** amphibian   **c.** bird   **d.** mammal

6. Reptiles get heat from ____.
   a. food   **b.** the sun   **c.** shade   **d.** a pouch

7. A crocodile is a kind of ____.
   a. amphibian   **b.** mammal   **c.** bird   **d.** reptile

8. Which animal is a type of fish? ____
   a. salamander   **b.** bear   **c.** trout   **d.** flamingo

9. What kind of animal is warm-blooded? ____
   a. fish   **b.** amphibian   **c.** bird   **d.** reptile

10. A joey is a baby ____.
   a. salamander   **b.** kangaroo   **c.** bear   **d.** frog

C COMPLETION Complete the sentences. *(20 points: 4 points each)*

Example: Vertebrates are animals with ___backbones___.

11. A _____ baby kangaroo lives in its mother's pouch.

12. Warm-blooded animals make their own body _____.

13. A reptile moves from sun to shade to control its _____.

14. Cold-blooded vertebrates get body heat from their _____.

15. A baby kangaroo lives in its mother's pouch and drinks _____.

GATEWAY TO SCIENCE Assessment Book · Copyright © Thomson Heinle

33

D **SHORT ANSWERS** Look at the table. Answer the questions.
(20 points: 4 points each)

Warm-blooded Vertebrates	Cold-blooded Vertebrates
Mammals and birds are warm-blooded.	Fish, amphibians, and reptiles are cold-blooded.
Body temperature stays about the same.	Body temperature changes.
They make body heat from the food they eat.	They get body heat from their surroundings.
They have backbones.	They have backbones.

Example: What are some cold-blooded vertebrates?

_____ fish, amphibians, and reptiles _____

16. How are a mammal's body temperature and a fish's body temperature different?

_____

17. Which animal's body temperature would be cooler on a cold day, a bird's or a

reptile's? _____

18. How are an amphibian's body temperature and a reptile's body temperature the

same? _____

19. What body parts do all vertebrates have?

_____

20. What kinds of animals don't need to lie in the sun to stay warm?

_____

E **WRITING** Imagine that you are a kangaroo. Describe the beginning of your life
from the time you are born until you are nine months old. Where do you live? What
do you eat? Write a paragraph. *(20 points)*

_____

_____

_____

_____

_____

Grade

A TRUE/FALSE Write if the sentence is true (T) or false (F). If the sentence is false, change the underlined word or phrase to make it true. *(20 points: 4 points each)*

Example: **Tissues** are groups of organs that work together. __F__ ____systems____

1. The circulatory system carries food to every **cell**. ____ _____

2. The heart is an **organ** of the body. ____ _____

3. The human body has **four** organ systems. ____ _____

4. The heart's job is to **pump blood**. ____ _____

5. An organ is made of **two or more** tissues. ____ _____

B MULTIPLE CHOICE Choose the correct answer. *(20 points: 4 points each)*

Example: Tissues are made of the same kind of __b__.
  a. organs   **b.** cells   **c.** systems   **d.** tubes

6. Blood vessels are ____.
  a. large organs   **b.** organ systems   **c.** small tubes   **d.** wastes

7. The ____ is the main part of the circulatory system.
  a. skin   **b.** lung   **c.** kidney   **d.** heart

8. The ____ take air in and out of the body.
  a. muscles   **b.** intestines   **c.** lungs   **d.** blood vessels

9. The excretory system is responsible for ____.
  a. removing wastes   **c.** protecting the body
  b. circulating blood   **d.** supporting the body

10. The right side of the heart pumps blood to the ____.
  a. brain   **b.** lungs   **c.** liver   **d.** stomach

C COMPLETION Complete the sentences. *(20 points: 4 points each)*

Example: Blood moves through the body in ____blood vessels____.

11. The _____ side of the heart pumps blood to the body.

12. The word _____ means "round."

13. _____ are the organs that support the body.

14. The _____ system is the body's transport system.

15. The _____ is an organ that protects the body.

Name _____   Date _____

D **SHORT ANSWERS** Look at the concept map. Answer the questions.
(20 points: 4 points each)

Four Organ Systems

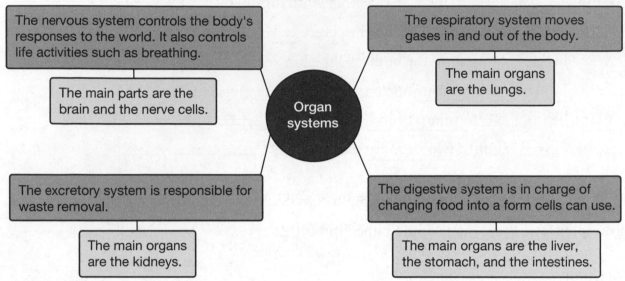

The nervous system controls the body's responses to the world. It also controls life activities such as breathing.

The main parts are the brain and the nerve cells.

The respiratory system moves gases in and out of the body.

The main organs are the lungs.

Organ systems

The excretory system is responsible for waste removal.

The main organs are the kidneys.

The digestive system is in charge of changing food into a form cells can use.

The main organs are the liver, the stomach, and the intestines.

Example: What job does the nervous system do?

  controls the body's responses to the world

16. What are the main parts of the excretory system?

  _____

17. What is the main job of the respiratory system?

  _____

18. What are the main organs in the respiratory system?

  _____

19. What organ system changes food into a form the body can use?

  _____

20. What are the main parts of the nervous system?

  _____

E **WRITING** Choose one organ system of the human body. Name its parts. Explain what it does. Write a paragraph. *(20 points)*

_____

_____

_____

_____

_____

Name _____   Date _____

Grade

A **TRUE/FALSE** Write if the sentence is true (T) or false (F). If the sentence is false, change the underlined word or phrase to make it true. *(20 points: 4 points each)*

Example: Some **bacteria** make new cells by vegetative reproduction. ___F___ ___plants___

1. **An onion** can grow a new bulb. ____ _____

2. Some plants can reproduce from their **stems**. ____ _____

3. The word part *phase* means **"end."** ____ _____

4. **All living things** use mitosis to reproduce. ____ _____

5. **Cellular reproduction** occurs in a series of stages. ____ _____

B **MULTIPLE CHOICE** Choose the correct answer. *(20 points: 4 points each)*

Example: In mitosis, a parent cell divides into two __c__.
   a. chromosomes   **b.** bacteria   **c.** daughter cells   **d.** bulbs

6. Prophase is ____ stage of mitosis.
   a. the first   **b.** the last   **c.** the parent   **d.** the vegetative

7. A strawberry plant uses ____ to reproduce.
   a. bulbs   **b.** runners   **c.** yeast   **d.** amoebas

8. In the second stage of mitosis, the chromosomes move ___.
   a. to the middle of the cell   **c.** away from each other
   b. along the ground   **d.** to the roots

9. Bacteria cells can make new cells in a few ____.
   a. days   **b.** hours   **c.** years   **d.** minutes

10. After mitosis, the daughter cells are ____ the parent cells.
   a. different from   **b.** before   **c.** the same as   **d.** around

C **COMPLETION** Complete the sentences. *(20 points: 4 points each)*

Example: In the first stage of mitosis, the wall around the ____nucleus____ goes away.

11. A kalanchoe plant grows new plants along its _____.

12. In anaphase, the third stage of mitosis, the _____ pairs move away from each other.

13. New strawberry plants grow where _____ touch the ground.

14. Chromosomes provide the instructions to make new _____.

15. In the telophase stage of _____, two new cells form.

Name _____ Date _____

D **SHORT ANSWERS** Look at the table. Answer the questions.
(20 points: 4 points each)

Bacteria Reproduce

Minutes	Number of Bacteria
0	1
15	2
30	4
45	8

Example: How often does a bacteria cell divide?

_____every 15 minutes_____

16. How many more bacteria cells are there after 30 minutes than after 15 minutes?

_____

17. How many more bacteria cells are there after 45 minutes than after 30 minutes?

_____

18. What happens to the number of bacteria cells every 15 minutes?

_____

19. How many bacteria cells will there be after 60 minutes has passed?

_____

20. After how many minutes will there be 32 bacteria cells?

_____

E **WRITING** You planted some onions and some strawberries. How can each kind of plant use vegetative reproduction to make new plants? Write a paragraph.
(20 points)

_____

_____

_____

_____

_____

GATEWAY TO SCIENCE Assessment Book • Copyright © Thomson Heinle

Name _____ Date _____

☐ Grade

A **TRUE/FALSE** Write if the sentence is true (T) or false (F). If the sentence is false, change the <u>underlined</u> word or phrase to make it true. *(20 points: 4 points each)*

Example: In meiosis, the sperm cell comes from the **mother**. __F__ _____father_____

1. Offspring of asexual reproduction are **different from** their parent. ____

   _____

2. In meiosis, cell division happens **two** times. ____ _____

3. A fertilized cell has some chromosomes from **each parent**. ____ _____

4. Chromosomes can **cross over** during meiosis. ____ _____

5. Offspring of sexual reproduction have **one parent**. ____ _____

B **MULTIPLE CHOICE** Choose the correct answer. *(20 points: 4 points each)*

Example: At the beginning of meiosis, pairs of __b__ line up.
   a. sex cells   **b.** chromosomes   **c.** eggs   **d.** daughter cells

6. Sexual reproduction can cause changes in ___.
   a. parent cells   **b.** line ups   **c.** flow charts   **d.** chromosomes

7. During crossing over, small sections of chromosomes ____.
   a. line up   **c.** switch places
   b. divide into daughter cells   **d.** move apart

8. *Line up* means ____.
   a. get into a line   **c.** divide chromosomes
   b. make sperm cells   **d.** variations happen

9. The word part *a* in *asexual* means ____.
   a. up   **b.** again   **c.** off   **d.** without

10. In meiosis, a cell becomes a set of four new ____.
   a. chromosomes   **b.** asexual cells   **c.** sex cells   **d.** fertilized cells

C **COMPLETION** Complete the sentences. *(20 points: 4 points each)*

Example: Chromosomes are in a cell's _____nucleus_____.

11. Many plants and animals reproduce by _____ reproduction.

12. During _____, chromosomes in the daughter cells become different

   from chromosomes in the parent cell.

13. If a parent cell comes from a female, meiosis ends with _____ cell.

14. In _____, every offspring has chromosomes from two parents.

15. Words such as *first, then,* and *finally* show the _____, or sequence, of events.

D SHORT ANSWERS Look at the flow chart. Answer the questions.
(20 points: 4 points each)

Reproduction

Sperm cell (from father)   Egg cell (from mother)   Fertilized cell (offspring)

Example: How many cells start this process?

_____2_____

16. How many parents does the offspring have?

_____

17. What kind of reproduction is shown?

_____

18. How many possible results does this flow chart have?

_____

19. What kind of cell comes from the mother?

_____

20. What is another name for the fertilized cell?

_____

E WRITING Your dog has puppies. They show variation in their markings. Explain why they look different from one another. Write a paragraph. *(20 points)*

_____

_____

_____

_____

_____

A TRUE/FALSE Write if the sentence is true (T) or false (F). If the sentence is false, change the <u>underlined</u> word or phrase to make it true. *(20 points: 4 points each)*

Example: <u>Genes</u> are made of DNA. __F__ __chromosomes__

1. **<u>Thymine</u>** provides the chemical recipe for traits. ____ _____

2. Offspring inherit pairs of **<u>chromosomes</u>** from their parents. ____ _____

3. **<u>Guanine controls</u>** traits in the body. ____ _____

4. Parents pass on traits in their **<u>bases</u>**. ____ _____

5. A Punnett square shows parents' **<u>chromosomes</u>**. ____ _____

B MULTIPLE CHOICE Choose the correct answer. *(20 points: 4 points each)*

Example: DNA is in the shape of __c__.

  a. guanine   **b.** a dominant gene   **c.** a double helix   **d.** a nucleus

6. Dominant and ____ have opposite meanings.

  a. twisted   **b.** recessive   **c.** square   **d.** control

7. Each gene in a body controls a certain ____.

  a. trait   **b.** nucleus   **c.** base   **d.** chromosome

8. A trait appears whenever a ____ gene is present.

  a. recessive   **b.** different   **c.** dominant   **d.** twisted

9. Guanine is a kind of ____.

  a. DNA   **b.** chromosome   **c.** trait   **d.** base

10. A base is a kind of ____.

  a. gene   **c.** double helix
  b. chemical compound   **d.** recipe

C COMPLETION Complete the sentences. *(20 points: 4 points each)*

Example: Adenine, thymine, guanine, and cytosine are the four ___bases___

  in DNA.

11. Chromosomes are in the _____ of a cell.

12. Eye color and hair color are _____.

13. DNA contains _____ on how to build a body.

14. Some traits appear only when both genes in a pair are _____.

15. A double helix looks like a twisted _____.

D **SHORT ANSWERS** Look at the Punnett square. Answer the questions.
(20 points: 4 points each)

Traits of Pea Plants
(tall (T) is the dominant trait)

Parent A

	T	t
T	TT	Tt
t	Tt	tt

Parent B

Example: What kind of letter in a Punnett square stands for a dominant gene?

_____*capital or uppercase*_____

16. What kind of letter in a Punnett square stands for a recessive gene?

_____

17. How many dominant and recessive genes does Parent A have?

_____

18. How many of the gene pairs in this Punnett square have two recessive genes?

_____

19. How many of the gene pairs in this Punnett square have two dominant genes?

_____

20. How many of the possible offspring gene pairs will show the dominant trait?

_____

E **WRITING** Describe DNA. What does DNA look like? What is it made of? What does it do? Write a paragraph. *(20 points)*

_____

_____

_____

_____

_____

Name _____ Date _____

Grade

A **TRUE/FALSE** Write if the sentence is true (T) or false (F). If the sentence is false, change the underlined word or phrase to make it true. *(20 points: 4 points each)*

Example: The shape of a duck's beak helps the duck **swim**. __F__ ____find food____

1. Birds of the same species **cannot have** variations. ____ _____

2. Evolution says that new species develop from **earlier species**. ____ _____

3. Birds are **organisms**. ____ _____

4. **An adaptation** helps an organism survive. ____ _____

5. Birds of different species **cannot reproduce** together. ____ _____

B **MULTIPLE CHOICE** Choose the correct answer. *(20 points: 4 points each)*

Example: In science, __c__ explains how something happens.
  a. a selection **b.** an adaptation **c.** a theory **d.** a species

6. Over time, ____ can occur in species.
  a. diagrams **b.** evidence **c.** theories **d.** variations

7. Organisms that once lived but don't exist today are ____.
  a. birds **b.** variations **c.** subspecies **d.** extinct

8. A ____ beak can help a heron catch fish.
  a. short, thick **b.** long, thin **c.** crossed **d.** common

9. Sometimes a variation becomes ____.
  a. a fossil **b.** an adaptation **c.** a subspecies **d.** an organism

10. About 200 million years ago, there were ____ on Earth.
  a. dinosaurs **b.** ducks **c.** swans **d.** sparrows

C **COMPLETION** Complete the sentences. *(20 points: 4 points each)*

Example: A duck's beak helping it find food is an example of ____adaptation____.

11. Birds of different _____ have variations, but they can still reproduce with one another.

12. Birds that can't reproduce together are members of different _____.

13. Variations become adaptations though the process of _____.

14. Birds see and eat dark moths on light _____.

15. Short, strong beaks help some grosbeaks eat _____.

Name _____   Date _____

D **SHORT ANSWERS** Look at the tree diagram. Answer the questions.
(20 points: 4 points each)

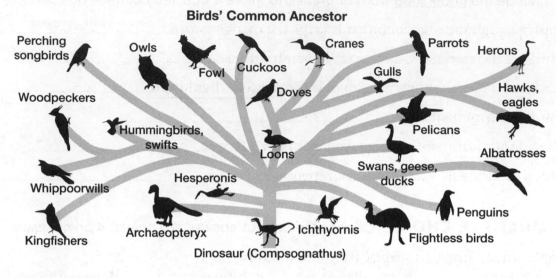

Birds' Common Ancestor

Example: What kind of birds are closely related to hawks and eagles?

_____ herons _____

16. Are penguins more closely related to albatrosses or to pelicans?

_____

17. What kind of bird is most closely related to a parrot?

_____

18. What kind of birds did doves evolve from?

_____

19. What was the first birdlike species to evolve from a dinosaur?

_____

20. Which bird is most closely related to fowl?

_____

E **WRITING** Explain how adaptations in birds' beaks help them find food. Give at least one example of how a beak helps a bird get and eat seeds. Give at least one example of how a beak helps a bird catch and eat fish. Write a paragraph. *(20 points)*

_____

_____

_____

_____

_____

GATEWAY TO SCIENCE Assessment Book • Copyright © Thomson Heinle

Name _____ Date _____

Student book pages 078–081

A **TRUE/FALSE** Write if the sentence is true (T) or false (F). If the sentence is false, change the <u>underlined</u> word or phrase to make it true. *(20 points: 4 points each)*

Example: A phylum is a group that is **larger** than a kingdom. ___F___ _____smaller_____

1. In binomial nomenclature an organism has **three names.** ____ _____

2. Seaweeds and protozoans belong to the **protist kingdom.** ____ _____

3. All **mushrooms** have a stem and a cap. ____ _____

4. Members of the protist kingdom live in **dry places.** ____ _____

5. Scientists use **DNA** to classify living organisms. ____ _____

B **MULTIPLE CHOICE** Choose the correct answer. *(20 points: 4 points each)*

Example: Closely related species have DNA that is __b__.
   a. smaller   **b.** similar   **c.** binomial   **d.** very different

6. A fern is a member of the ____.
   a. animal kingdom   **b.** protist kingdom   **c.** plant kingdom   **d.** cheetah family

7. Carolus Linnaeus created ____.
   a. DNA   **c.** instructions on how to make new cells
   b. a diagram   **d.** a classification system

8. Mushrooms belong to the ____.
   a. eubacteria kingdom   **c.** plant kingdom
   b. fungi kingdom   **d.** order or carnivores

9. Halophiles live in ____.
   a. salty water   **c.** animals' stomachs
   b. dry places   **d.** pine trees

10. Seaweeds are ____ protists.
   a. many-celled   **b.** one-celled   **c.** two-part   **d.** rod-shaped

C **COMPLETION** Complete the sentences. *(20 points: 4 points each)*

Example: Rod-shaped bacteria are in the eubacteria _____kingdom_____.

11. In binomial nomenclature, the second part is the _____.

12. Molds are part of the _____ kingdom.

13. Methanogens are bacteria that live in animals' _____.

14. *Panthera pardus* and *Panthera onca* belong to the same _____.

15. Scientists classify living organisms into six _____.

D SHORT ANSWERS Look at the diagram. Answer the questions.
(20 points: 4 points each)

Classifying a Lion

Fly	Bird	Mouse	Dog	Cheetah	Tiger	Lion	Kingdom: Animal

Phylum: Chordate

Class: Mammal

Order: Carnivore

Family: Felidae

Genus: *Panthera*

Species: *Panthera leo*

Example: What phylum does a mouse belong to?

_____ *chordate* _____

16. What class is a dog a part of? _____

17. What three animals are members of the family *felidae*?

_____

18. Dogs and lions are members of what order?

_____

19. Which two animals are not mammals?

_____

20. Which animal is not a chordate? _____

E WRITING Who was Carolus Linnaeus? What did he create? What is it called? How do we use it? Write a paragraph. *(20 points)*

_____

_____

_____

_____

_____

Classification Systems   LIFE SCIENCE

GATEWAY TO SCIENCE Assessment Book • Copyright © Thomson Heinle

Name _____ Date _____

Grade

A **TRUE/FALSE** Write if the sentence is true (T) or false (F). If the sentence is false, change the <u>underlined</u> word or phrase to make it true. *(20 points: 4 points each)*

Example: A rain forest is a kind of community. ___F___ ___biome___

1. A <u>**habitat**</u> is a large area with a certain temperature and rainfall. ____ _____

2. <u>**Deciduous forests**</u> have four seasons. ____ _____

3. A <u>**biome**</u> is part of an ecosystem. ____ _____

4. The weather is very <u>**dry**</u> in a rain forest. ____ _____

5. Cactus plants <u>**are equipped**</u> to live in a desert. ____ _____

B **MULTIPLE CHOICE** Choose the correct answer. *(20 points: 4 points each)*

Example: It is ___d___ in the tundra.
   a. hot   **b.** rainy   **c.** wet   **d.** cold

6. ____ gets less than 25 cm of rain in a year.
   a. A forest   **b.** An ecosystem   **c.** A desert   **d.** A pond

7. The word part *eco* in *ecosystem* means ____.
   a. soil   **b.** house   **c.** plant   **d.** species

8. A ____ includes all the members of a species in an area.
   a. biome   **b.** community   **c.** forest   **d.** population

9. A habitat is an animal's ____.
   a. home   **b.** ecosystem   **c.** biome   **d.** community

10. A ____ is an area where plants and animals live.
   a. population   **b.** burrow   **c.** biome   **d.** habitat

C **COMPLETION** Complete the sentences. *(20 points: 4 points each)*

Example: Water, rocks, and soil are _____nonliving_____ things in an ecosystem.

11. Deciduous trees _____ every fall.

12. Cactus plants have special places for keeping _____.

13. A _____ shows things that take place over time.

14. The climate in a _____ is hot and dry.

15. Each kind of living thing in an ecosystem is a _____.

D **SHORT ANSWERS** Look at the timeline. Answer the questions.
(20 points: 4 points each)

Ecological Succession: From Pond to Forest

1870	1900	1930	1990 Today

Pond filling in, no grass   Soil   Grasses, shrubs, and young trees   Forest with large trees

Example: Were grasses growing in the pond in 1870?

_____*There were no grasses in 1870*_____.

16. When did soil start to fill the pond? _____

17. Which grew in the pond first, grasses or large trees?

_____

18. When did shrubs start to grow? _____

19. When did the place become a forest?

_____

20. How long did it take the pond community to become a forest community?

_____

E **WRITING** Think of a kangaroo rat. Describe its habitat. How does it get the things it needs? Write a paragraph. *(20 points)*

_____

_____

_____

_____

_____

GATEWAY TO SCIENCE Assessment Book · Copyright © Thomson Heinle

Name _____ Date _____

Grade

A **TRUE/FALSE** Write if the sentence is true (T) or false (F). If the sentence is false, change the underlined word to make it true. *(20 points: 4 points each)*

Example: <u>Animals</u> make food energy from sunlight. __F__ _____plants_____

1. A bobcat catches fish as **prey**. ____ _____

2. Grasshoppers are **producers**. ____ _____

3. **Bacteria** break down dead plants and animals for energy. ____ _____

4. A **producer** hunts and eats prey. ____ _____

5. A frog is a **consumer**. ____ _____

B **MULTIPLE CHOICE** Choose the correct answer. *(20 points: 4 points each)*

Example: Some bees live in __c__ with certain flowering plants.
a. a food chain   c. symbiosis
b. an energy pyramid   d. photosynthesis

6. Bacteria and fungi are ____.
a. predators   b. decomposers   c. producers   d. prey

7. Energy moves from plant to animal in ____.
a. a food chain   b. symbiosis   c. mutualism   d. the intestines

8. In commensalism, two kinds of organisms live together. One animal is ____ and the other is not harmed or helped.
a. hunted   b. consumed   c. harmed   d. helped

9. The word part *herb* in *herbivore* means ____.
a. meat   b. prey   c. predator   d. grass

10. A predator hunts and eats ____.
a. animals   b. plants   c. fungi   d. producers

C **COMPLETION** Complete the sentences. *(20 points: 4 points each)*

Example: A tapeworm lives in another animal's _____intestines_____.

11. In parasitism, one organism in the relationship is harmed and the other is

_____.

12. Some prey have coloring that helps them _____ from predators.

13. Many different food chains in one area make up a _____.

14. An animal that eats other plants and animals to get energy is a _____.

15. Not all energy _____ to the next level in an energy pyramid.

D **SHORT ANSWERS** Look at the energy pyramid. Answer the questions.
(20 points: 4 points each)

An Energy Pyramid

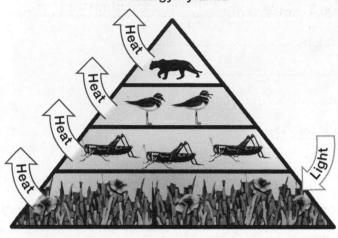

Example: Are producers at the top or the bottom of the energy pyramid?

_____the bottom_____

16. What do producers use to get energy?

_____

17. Are birds producers or consumers? _____

18. What is given off at each level of the pyramid?

_____

19. Is a tiger at the top or the bottom or a food chain?

_____

20. Why is less energy passed on to each level of the pyramid?

_____

E **WRITING** An owl gets energy by catching and eating a snake. Where did the snake get energy? Describe a possible food chain. Use the words *predator* and *prey*. Write a paragraph. *(20 points)*

_____

_____

_____

_____

_____

Name _____ Date _____

Student book pages 090–093

Grade

A **TRUE/FALSE** Write if the sentence is true (T) or false (F). If the sentence is false, change the underlined word or phrase to make it true. *(20 points: 4 points each)*

Example: Evaporation is the process of **condensing**. __F__ ____evaporating____

1. **Bacteria** in soil can use nitrogen gas. ____ _____

2. Animals give off **oxygen**. ____ _____

3. Living things discharge **gases**. ____ _____

4. Plants take in water through their **stems**. ____ _____

5. Animals breathe in **carbon dioxide**. ____ _____

B **MULTIPLE CHOICE** Choose the correct answer. *(20 points: 4 points each)*

Example: Water changes to water vapor during __b__.
  a. precipitation   **b.** evaporation   **c.** condensation   **d.** runoff

6. Air is mostly ____.
  a. water vapor   **b.** carbon dioxide   **c.** nitrogen   **d.** oxygen

7. Plants discharge ____ through their leaves.
  a. water vapor   **b.** clouds   **c.** rain   **d.** nitrogen

8. Most water evaporates from ____.
  a. clouds   **b.** plants   **c.** animals   **d.** the ocean

9. Animal wastes contain ____.
  a. nitrogen   **b.** oxygen   **c.** water vapor   **d.** carbon dioxide

10. Clouds form when ____ happens.
  a. evaporation   **b.** transpiration   **c.** condensation   **d.** runoff

C **COMPLETION** Complete the sentences. *(20 points: 4 points each)*

Example: Condensation happens when water vapor ____cools____.

11. Plants take in nitrogen from _____.

12. Plants use _____ from the air.

13. When water droplets in clouds become large enough, they _____ as precipitation.

14. Most living things cannot use the _____ in the air.

15. When rain goes into the ground, it becomes _____.

D SHORT ANSWERS Look at the model. Answer the questions.
(20 points: 4 points each)

The Nitrogen Cycle

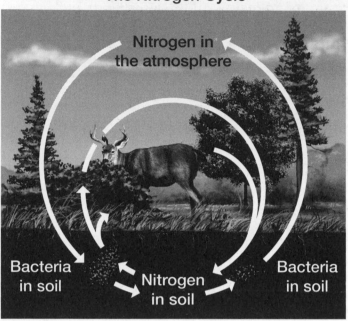

Example: What does the deer get from the plants it eats?
___nitrogen___

16. Where does nitrogen in the soil come from?
_____

17. What do soil bacteria get from the atmosphere?
_____

18. Where does the nitrogen in the atmosphere come from?
_____

19. To what two places do bacteria release nitrogen?
_____

20. What do the arrows in the model show?
_____

E WRITING You have a carrot plant in your kitchen: What does the plant do to keep you alive? What do you do to keep the plant alive? Write a paragraph. *(20 points)*

_____

_____

_____

_____

_____

A **TRUE/FALSE** Write if the sentence is true (T) or false (F). If the sentence is false, change the <u>underlined</u> word or phrase to make it true. *(20 points: 4 points each)*

Example: Plant **roots** respond to light. __F__   leaves and flowers

1. A change in seasons can be **an instinct**. ____ _____

2. Caribou migrate north **in winter**. ____ _____

3. You do a **reflex** without thinking about it. ____ _____

4. A **loud noise** can be a stimulus. ____ _____

5. Squinting is a **learned behavior**. ____ _____

B **MULTIPLE CHOICE** Choose the correct answer. *(20 points: 4 points each)*

Example: A plant growing toward light is showing __a__.
   a. phototropism   **c.** migration
   b. gravitropism   **d.** learned behavior

6. Plant roots respond to ____.
   a. noise   **b.** light   **c.** gravity   **d.** insects

7. Some animals estivate to escape ____.
   a. cold weather   **b.** loud noises   **c.** bright lights   **d.** hot weather

8. Hibernation is a ____.
   a. migration   **c.** response to gravity
   b. deep sleep   **d.** stimulus

9. Plant stems grow ____ in response to gravity.
   a. downward   **b.** north   **c.** slowly   **d.** upward

10. Migration is ____ in birds.
   a. a learned behavior   **c.** an instinct
   b. a stimulus   **d.** a loud noise

C **COMPLETION** Complete the sentences. *(20 points: 4 points each)*

Example: Many birds fly _____south_____ for the winter.

11. Some _____ open only in the morning.

12. Putting on sunglasses to protect your eyes is a _____.

13. A missing egg is a _____ to a graylag goose.

14. Sundew plants have sticky leaves that attract _____.

15. A chipmunk _____ underground in winter.

Name _____ Date _____

D **SHORT ANSWERS** Look at the drawings. Answer the questions.
(20 points: 4 points each)

A: Venus Fly Trap.   B: A plant's root and stem.

Example: What is in the leaf of the Venus fly trap in Drawing A?

_____ *an insect* _____

16. What is the Venus fly trap's leaf doing?

_____

17. What does the Venus fly trap use the insect for?

_____

18. What movement will the fly trap's leaf make after it uses the insect?

_____

19. What causes the root of the plant in Drawing B to grow down?

_____

20. What does phototropism cause the leaves to do in Drawing B?

_____

E **WRITING** Describe how a chipmunk and a frog respond to weather. What do
they do? When do they do it? What are their behaviors called? Tell how their behaviors are the same. Write a paragraph. *(20 points)*

_____

_____

_____

_____

_____

Name _____ Date _____

📖 Student book pages 098–101

Grade

A **TRUE/FALSE** Write if the sentence is true (T) or false (F). If the sentence is false, change the <u>underlined</u> word or phrase to make it true. *(20 points: 4 points each)*

Example: Endangered species exist in **large** numbers. __F__ _____small_____

1. The dodo is an **endangered** species. ____ _____

2. Some **subspecies** of tigers are extinct. ____ _____

3. People use some **plants** for medicines. ____ _____

4. Giant pandas **are extinct**. ____ _____

5. Some animals die out because of **climate change**. ____ _____

B **MULTIPLE CHOICE** Choose the correct answer. *(20 points: 4 points each)*

Example: Bald eagles were once __c__ in the United States.
   a. a subspecies   **b.** extinct   **c.** common   **d.** used as medicine

6. The prefix *en* in the word *endangered* means ____.
   a. harmful   **b.** extinct   **c.** weak   **d.** to cause to be

7. Bald eagles are ____ by the United States government.
   a. hunted   **b.** polluted   **c.** protected   **d.** endangered

8. The ____ is extinct.
   a. mammoth   **c.** black rhino
   b. humpback whale   **d.** whooping crane

9. Loss of ____ can cause a species to become extinct.
   a. pollution   **b.** habitat   **c.** medicine   **d.** chemicals

10. People ____ humpback whales almost to extinction.
   a. studied   **b.** protected   **c.** hunted   **d.** saved

C **COMPLETION** Complete the sentences. *(20 points: 4 points each)*

Example: Some animals are endangered because they cannot find enough
_____food or space_____.

11. During the 1950s, eagles ate polluted _____.

12. Scientists hope to save the plants in the _____.

13. The actions of _____ threaten many species.

14. DDT caused bald eagles' eggshells to _____.

15. All tigers belong to the same _____.

Name _____ Date _____

D **SHORT ANSWERS** Look at the map. Answer the questions.
(20 points: 4 points each)

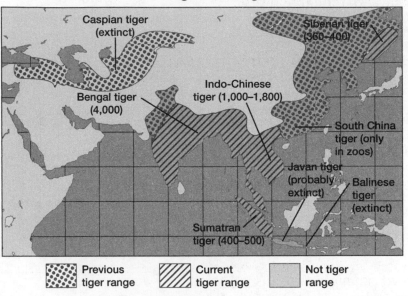

Tigers in Danger

Caspian tiger (extinct)

Siberian tiger (360–400)

Indo-Chinese tiger (1,000–1,800)

Bengal tiger (4,000)

South China tiger (only in zoos)

Javan tiger (probably extinct)

Balinese tiger (extinct)

Sumatran tiger (400–500)

Previous tiger range

Current tiger range

Not tiger range

Example: Are tiger ranges today larger or smaller than they were before?
_____ smaller _____

16. Are Caspian tigers extinct or endangered?

_____

17. About how many Bengal tigers exist?

_____

18. Do more Siberian tigers or Sumatran tigers exist?

_____

19. Which subspecies of tiger has the smallest population today?

_____

20. How many subspecies of tigers have become extinct?

_____

E **WRITING** Scientists are trying to save plants in the rain forests. Explain why. How do they think they can use the plants? Write a paragraph. *(20 points)*

_____

_____

_____

_____

GATEWAY TO SCIENCE Assessment Book • Copyright © Thomson Heinle

Grade

A TRUE/FALSE Write if the sentence is true (T) or false (F). If the sentence is false, change the underlined word or phrase to make it true. *(20 points: 2 points each)*

Example: Ribosomes store water, food, and waste. __F__ _____vacuoles_____

1. **Lysosomes** break down material. ____ _____

2. The lungs and the trachea belong to the **respiratory system**. ____ _____

3. Conifers are **seedless** plants. ____ _____

4. Animals **reproduce** to make new baby animals. ____ _____

5. The body temperature of **cold-blooded** animals always is the same.

   ____ _____

6. **An onion** can grow a new bulb. ____ _____

7. **Cytosine** provides the chemical recipe for traits. ____ _____

8. In binomial nomenclature an organism has **one name**. ____ _____

9. A bobcat catches fish as **prey**. ____ _____

10. A change in seasons can be **an instinct**. ____ _____

B MULTIPLE CHOICE Choose the correct answer. *(40 points: 2 points each)*

Example: The __b__ packages proteins.
   a. cytoplasm   **b.** Golgi complex   **c.** cell wall   **d.** mitochondria

11. Robert Hooke saw ____.
   a. the walls of a cell   **c.** the organelles of a cell
   b. the nucleus of a cell   **d.** a Golgi complex

12. A trichonympha is a kind of ____.
   a. protozoan   **b.** fungus   **c.** pseudopod   **d.** tube worm

13. In cell division, an organism ____.
   a. collects gases   **c.** makes new cells
   b. hunts and eats bacteria   **d.** helps grow new bones

14. Leaves use energy from ____ to make food.
   a. roots   **b.** sunlight   **c.** soil   **d.** flowers

B **MULTIPLE CHOICE,** continued

15. A flowering plant's ___ contains eggs.

   a. anther   **b.** stem   **c.** roots   **d.** pistil

16. The word part *photo* in *photosynthesis* means ____.

   a. to grow   **b.** oxygen   **c.** together   **d.** light

17. Animals such as ____ have tough outer shells.

   a. bobcats   **b.** owls   **c.** worms   **d.** insects

18. A spider is a type of ____.

   a. sea star   **b.** jellyfish   **c.** beetle   **d.** arachnid

19. Reptiles get heat from ____.

   a. water   **b.** the sun   **c.** shade   **d.** a pouch

20. Blood vessels are ____.

   a. large organs   **b.** organ systems   **c.** tubes   **d.** wastes

21. Prophase is ____ stage of mitosis.

   a. the first   **b.** the last   **c.** the parent   **d.** the sex cell

22. Sexual reproduction can cause changes in ____.

   a. parent cells   **b.** line-ups   **c.** the process   **d.** chromosomes

23. Dominant and ____ have opposite meanings.

   a. twisted   **b.** recessive   **c.** square   **d.** variation

24. Over time, ____ can occur in species.

   a. traits   **b.** evidence   **c.** theories   **d.** variations

25. A fern is a member of the ____.

   a. animal kingdom   **c.** plant kingdom
   b. protist kingdom   **d.** fungus family

26. ____ gets less than 25 cm of rain in a year.

   a. A rain forest   **b.** An ecosystem   **c.** A desert   **d.** A pond

Name _____ Date _____

B MULTIPLE CHOICE, continued

27. Bacteria and fungi are ____.
    a. predators   b. decomposers   c. producers   d. eubacteria

28. Air is mostly ____.
    a. water vapor   c. nitrogen
    b. carbon monoxide   d. oxygen

29. Plant roots respond to ____.
    a. noise   b. light   c. gravity   d. worms

30. The prefix *en* in the word *endangered* means ___.
    a. strong   b. extinct   c. weak   d. to cause to be

C COMPLETION Complete the sentences. *(20 points: 2 points each)*

Example: There are rod, spiral, and _____round_____ bacteria.

31. Yeast is a one-celled kind of _____.

32. Humus has lots of _____ that plants need to grow.

33. Wide rings in a tree trunk mean a year with a lot of _____.

34. A sponge's body is _____ at the bottom.

35. The _____ side of the heart pumps blood out to the body.

36. Many plants and animals reproduce by _____ reproduction.

37. Birds of different _____ have variations, and they can still reproduce with one another.

38. Deciduous trees lose their _____ every fall.

39. Plants take nitrogen from the _____.

40. During the 1950s, many eagles ate polluted _____.

D **SHORT ANSWERS** Look at the diagram. Answer the questions.
(20 points: 2 points each)

Classifying a Lion

| | | | | | | | Kingdom: Animal |
| Fly | Bird | Mouse | Dog | Cheetah | Tiger | Lion | |

Phylum: Chordate

Class: Mammal

Order: Carnivore

Family: Felidae

Genus: *Panthera*

Species: *Panthera leo*

Example: What phylum does a mouse belong to?

_____ chordate _____

41. What class does a dog belong to?

_____

42. Which three animals are members of the family *felidae*?

_____

43. Dogs and lions are members of which order?

_____

44. What two animals are not mammals?

_____

45. What animal is not a chordate?

_____

GATEWAY TO SCIENCE Assessment Book • Copyright © Thomson Heinle

D **SHORT ANSWERS,** continued

Look at the table. Answer the questions.

Bacteria Reproduction

Minutes	Number of Bacteria
0	1
15	2
30	4
45	8

46. How many more bacteria cells are there after 30 minutes than there are after
15 minutes? _____

47. How many more bacteria cells are there after 45 minutes than there are after
30 minutes? _____

48. What happens to the number of bacteria cells every 15 minutes?
_____

49. How many bacteria cells will there be after 60 minutes have passed?
_____

50. After how many minutes will there be 32 bacteria cells?
_____

E **WRITING ASSESSMENT** Write paragraphs. *(100 points: 25 points each)*

51. You grow two plants. One is a fern, and the other one is a moss. How are they alike? How are they different?

_____

_____

_____

_____

_____

52. Imagine that you are a drop of blood. Explain the trip you take from the heart, through the body, and back to the heart. Where do you leave the heart? Where do you go after that? How do you travel?

_____

_____

_____

_____

_____

53. What are species? What are subspecies? How are they different? Give some examples.

_____

_____

_____

_____

54. When did bald eagles become endangered? How did it happen? Why? Tell what humans did to cause the problem. Tell what the eagles did. What happened after that?

_____

_____

_____

_____

Grade ☐

A **TRUE/FALSE** Write if the sentence is true (T) or false (F). If the sentence is false, change the <u>underlined</u> word or phrase to make it true. *(20 points: 4 points each)*

Example: Space is the area beyond **the sun**. __F__ _____Earth_____

1. There are **hundreds** of galaxies in the universe. ____ _____

2. The Milky Way contains about 100 billion **galaxies**. ____ _____

3. Scientists use telescopes to study the **planets**. ____ _____

4. Proxima Centauri is a **spiral galaxy**. ____ _____

5. Distances in space are very **small**. ____ _____

B **MULTIPLE CHOICE** Choose the correct answer. *(20 points: 4 points each)*

Example: Galaxies are the __c__ objects in space.
- **a.** smallest
- **b.** nearest
- **c.** largest
- **d.** emptiest

6. The Milky Way is a ____.
- **a.** star
- **b.** planet
- **c.** galaxy
- **d.** telescope

7. Some telescopes collect ____ with lenses.
- **a.** light waves
- **b.** objects
- **c.** light-years
- **d.** gases

8. A spiral galaxy has ____.
- **a.** mirrors
- **b.** lenses
- **c.** arms
- **d.** dishes

9. The word part *tele* in *telescope* means ____.
- **a.** light-year
- **b.** radio
- **c.** far
- **d.** bulge

10. New stars form near the center of a ____.
- **a.** solar system
- **b.** sun
- **c.** planet
- **d.** galaxy

C **COMPLETION** Complete the sentences. *(20 points: 4 points each)*

Example: The Milky Way is about 100,000 _____light years_____ from end to end.

11. Radio telescopes collect _____ in bowl-shaped dishes.

12. A light-year is how far light travels _____.

13. The distance from Earth to the sun is one _____.

14. Some telescopes collect light waves with lenses and _____.

15. A light-year equals 9.46 trillion _____.

D **SHORT ANSWERS** Look at the table. Answer the questions.
(20 points: 4 points each)

Distances in the Solar System	
Planet	**Distance From the Sun (AU)**
Mercury	0.39
Venus	0.72
Earth	1
Mars	1.5
Jupiter	5.2
Saturn	9.5
Uranus	19.2
Neptune	30.1

Example: Which planet is closest to Neptune?

_____Uranus_____

16. Which planet is closest to Earth? _____

17. How far is Mars from Earth? _____

18. How far from the sun is Saturn? _____

19. Which planet is 19.2 AU from the sun?

_____

20. Which planet is closer to Earth: Jupiter or Mercury?

_____

E **WRITING** Describe telescopes and radio telescopes. What are their parts? How do they work? How are telescopes and radio telescopes alike and different? Write a paragraph. *(20 points)*

_____

_____

_____

_____

_____

GATEWAY TO SCIENCE Assessment Book • Copyright © Thomson Heinle

GATEWAY TO SCIENCE Assessment Book · Copyright © Thomson Heinle

Name _____ Date _____

Grade

A TRUE/FALSE Write if the sentence is true (T) or false (F). If the sentence is false, change the underlined word or phrase to make it true. *(20 points: 4 points each)*

Example: Red stars are **hotter** than the sun. __F__ _____cooler_____

1. Magnitude is a measure of **size**. ____ _____

2. A supergiant is **larger** than the sun. ____ _____

3. A **dim star** has a higher magnitude. ____ _____

4. Our sun is a **red** star. ____ _____

5. A dwarf star is very **large**. ____ _____

B MULTIPLE CHOICE Choose the correct answer. *(20 points: 4 points each)*

Example: A star is born in a __b__.
a. constellation   c. black hole
b. nebula   d. white dwarf

6. A main sequence star burns ____ as fuel.
a. dust   b. neutrons   c. hydrogen   d. gravity

7. The word part *magni* in *magnitude* means ____.
a. bright   b. dim   c. dwarf   d. big

8. Knowing ____ helps people find certain stars in the sky.
a. magnitudes   b. constellations   c. black holes   d. sizes

9. A smaller main sequence star can swell and become a ____.
a. red giant   b. neutron star   c. black hole   d. constellation

10. A supergiant might explode and become a ____.
a. main sequence star   c. supernova
b. a white dwarf   d. a nebula

C COMPLETION Complete the sentences. *(20 points: 4 points each)*

Example: Nothing can escape from a _____black hole_____.

11. When a white dwarf shrinks, it leaves behind a new _____.

12. A black hole has very strong _____.

13. A nebula is a cloud of _____.

14. A star's magnitude depends partly on how _____ it is to Earth.

15. Ursa Major and Ursa Minor are _____.

65

D **SHORT ANSWERS** Look at the Hertzsprung-Russell Diagram. Answer the questions. *(20 points: 4 points each)*

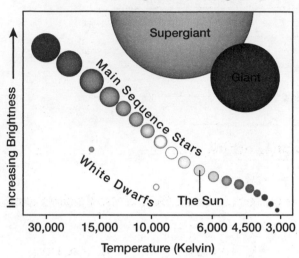

The Hertzsprung-Russell Diagram

Example: What do the numbers along the bottom of the diagram show?

_____temperature_____

16. Are the stars on the left of the diagram hotter or colder than the stars on the right?

_____

17. Are the stars at the bottom of the diagram bright or dim?

_____

18. Is a white dwarf hotter or colder than the sun?

_____

19. Is the sun brighter or dimmer than a giant?

_____

20. Where on the diagram would you find a cool, dim star?

_____

E **WRITING** Describe the life of a smaller main sequence star. Where is it born? What fuel does it burn? How does it change shape and color? How does its life end? Write a paragraph. *(20 points)*

_____

_____

_____

_____

_____

EARTH SCIENCE Stars

Grade

📖 Student book pages 110–113

A **TRUE/FALSE** Write if the sentence is true (T) or false (F). If the sentence is false, change the underlined word or phrase to make it true. *(20 points: 4 points each)*

Example: Jupiter is made of **metal**. __F__ ____gases____

1. A **dwarf planet** has a glowing tail. ____ _____

2. A **meteoroid** can be as small as dust. ____ _____

3. Eris and Pluto are **dwarf planets**. ____ _____

4. **Mars** is the largest planet. ____ _____

5. Comets revolve around the sun in **circular orbits**. ____ _____

B **MULTIPLE CHOICE** Choose the correct answer. *(20 points: 4 points each)*

Example: An orbit is __a__ or a track.
   a. a path   **b.** a planet   **c.** an axis   **d.** a season

6. Jupiter's Great Red Spot is a giant ____.
   a. comet   **b.** storm   **c.** asteroid   **d.** planet

7. A comet's glowing tail is made of ____.
   a. metal   **b.** rock   **c.** gas   **d.** ice

8. A dwarf planet orbits ____.
   a. Earth   **b.** Jupiter   **c.** the sun   **d.** an asteroid

9. Asteroids are large chunks of ____.
   a. rock and gases   **c.** ice and gases
   b. rock and metal   **d.** ice and metal

10. Our solar system has eight ____.
   a. planets   **b.** moons   **c.** comets   **d.** asteroids

C **COMPLETION** Complete the sentences. *(20 points: 4 points each)*

Example: Pluto and Eris appear to be made of ____rock and ice____.

11. Sir Edmund Halley learned about a famous _____.

12. Jupiter has about 60 _____.

13. Mercury is the planet closest to _____.

14. Asteroids and meteoroids orbit _____.

15. _____ appears about every 76 years.

Name _____ Date _____

D **SHORT ANSWERS** Look at the illustration. Answer the questions.
(20 points: 4 points each)

Jupiter and Its Largest Moons

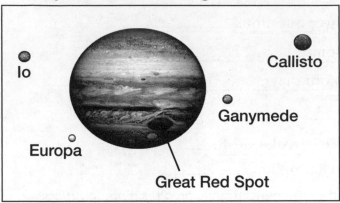

Io

Callisto

Ganymede

Europa

Great Red Spot

Example: What are the names of Jupiter's largest moons?

Io, Europa, Ganymede, and Callisto

16. How many moons are shown in the picture?

_____

17. Which moon looks farthest from Jupiter?

_____

18. Which of these moons looks the largest?

_____

19. What feature of Jupiter is shown, besides its moons?

_____

20. Does the Great Red Spot look larger or smaller than Jupiter's largest moon?

_____

E **WRITING** Describe Halley's comet. What is it made of? What makes it glow? Explain why it is named after Sir Edmund Halley. Write a paragraph. *(20 points)*

_____

_____

_____

_____

_____

GATEWAY TO SCIENCE Assessment Book • Copyright © Thomson Heinle

Name _____   Date _____

📖 Student book pages 114–117

Grade ☐

A **TRUE/FALSE** Write if the sentence is true (T) or false (F). If the sentence is false, change the <u>underlined</u> word or phrase to make it true. *(20 points: 4 points each)*

Example: Some trees lose their leaves in the **summer**. _F_ _____winter_____

1. Good scientists use data to **answer questions**. ____ _____

2. A year is divided into four **seasons**. ____ _____

3. The word *rotate* comes from a word that means "**to turn around**." ____
_____

4. Earth rotates on its **axis**. ____ _____

5. The part of Earth facing the sun has **night**. ____ _____

B **MULTIPLE CHOICE** Choose the correct answer. *(20 points: 4 points each)*

Example: Changes in the way the moon appears are called __c__.
   a. seasons   **b.** orbits   **c.** phases   **d.** nights

6. The warmest season on Earth is ____.
   a. spring   **b.** fall   **c.** summer   **d.** winter

7. On Earth, one ____ is 24 hours long.
   a. day and night   **b.** orbit   **c.** year   **d.** phase

8. The moon revolves, or ____ around Earth.
   a. tilts   **b.** orbits   **c.** rotates   **d.** spins

9. The word *appears* means the same as the word ____.
   a. orbits   **b.** seems   **c.** tilts   **d.** rotates

10. The part of Earth facing away from the sun has ____.
   a. summer   **b.** day   **c.** winter   **d.** night

C **COMPLETION** Complete the sentences. *(20 points: 4 points each)*

Example: It takes about _____365 days_____ for Earth to revolve around the sun.

11. The lengths of day and night change with the _____.

12. When the northern half of Earth has summer, the _____ has winter.

13. The _____ of the moon seems to change as it orbits Earth.

14. When the northern half of Earth tilts _____, it is winter there.

15. Earth's rotation makes _____ look like it is moving across the sky during the day.

Name _____   Date _____

D **SHORT ANSWERS** Look at the table. Answer the questions.
(20 points: 4 points each)

City	Length of day January 1	Length of day July 1
Nome, Alaska	4 hours, 12 minutes	21 hours, 9 minutes
Chicago, Illinois	9 hours, 13 minutes	15 hours, 11 minutes
Miami, Florida	10 hours, 34 minutes	13 hours, 43 minutes

Example: Which day is shortest in all cities?

_____ January 1 _____

16. Which city has the shortest day?

_____

17. Which city has the longest day?

_____

18. How many more hours of daylight does Chicago have on July 1 than on January 1?

_____

19. Which city has the most hours of daylight in January?

_____

20. Is the day longer in Chicago or in Miami on July 1?

_____

E **WRITING** Describe the movements of Earth in the solar system. How long does it take to rotate? How long does it take to revolve around the sun? Write a paragraph.
(20 points)

_____

_____

_____

_____

_____

GATEWAY TO SCIENCE Assessment Book • Copyright © Thomson Heinle

Name _____ Date _____

Grade

A TRUE/FALSE Write if the sentence is true (T) or false (F). If the sentence is false, change the underlined word or phrase to make it true. *(20 points: 4 points each)*

Example: The Latin word *lunaris* means "**sun**." __F__ _____moon_____

1. The Bay of Fundy has a **large** tidal range. ____ _____

2. The **sun** is the closest thing in space to Earth. ____ _____

3. The **lowest tide** happens during the moon's first- and third-quarter phases.

   ____ _____

4. The moon can cast a shadow on **Earth**. ____ _____

5. Tides rise and fall **once** each day. ____ _____

B MULTIPLE CHOICE Choose the correct answer. *(20 points: 4 points each)*

Example: Earth's shadow results in, or __d__, a lunar eclipse.
   a. blocks   **b.** rises   **c.** falls   **d.** causes

6. A ____ happens when the moon passes between the sun and Earth.
   a. lunar eclipse   **b.** solar eclipse   **c.** neap tide   **d.** tidal bulge

7. The sun, moon, and Earth line up during ____.
   a. a full moon   **c.** a tidal range
   b. a third-quarter moon   **d.** a neap tide

8. In a total solar eclipse, people on Earth see a glow around ____.
   a. the sun   **b.** Earth   **c.** the Bay of Fundy   **d.** the moon

9. When something blocks light, it causes ____.
   a. a tide   **b.** a bulge   **c.** a shadow   **d.** an ocean

10. Tides make sea levels ____.
   a. block sunlight   **b.** rise and fall   **c.** cause shadows   **d.** line up

C COMPLETION Complete the sentences. *(20 points: 4 points each)*

Example: The highest tide is called a _____spring tide_____.

11. The difference between high tide and low tide is called the _____.

12. Part of the sun's light is blocked in a _____ eclipse.

13. The sea level rises and falls a few _____ each day.

14. The sun, the moon, and Earth _____ during a new moon.

15. During a neap tide, the sun is at a _____ to the moon.

Name _____   Date _____

D **SHORT ANSWERS** Look at the photos. Answer the questions.
(20 points: 4 points each)

Photo A

Photo B

Tide on June 3 at 6:02 A.M. and 6:36 P.M.   Tide on June 3 at 12:41 P.M.

Example: You want to get on a boat on the morning of June 3. What time should you
go? _____ 6:02 A.M. _____

16. You want to walk on the beach on the afternoon of June 3. What time should
you go? _____

17. You want to get on a boat in the evening on June 3. What time should you go?
_____

18. Does Photo B show a high tide or a low tide?
_____

19. At about what time on June 3 did the first low tide happen?
_____

20. Why do you think there is no photo of the first low tide?
_____

E **WRITING** You are watching a total lunar eclipse. Where are the sun, the moon,
and Earth? What happens to the moon? What causes this? Describe what you see.
Write a paragraph. *(20 points)*

_____

_____

_____

_____

_____

Name _____ Date _____

☐ Grade

📖 Student book pages 122–125

A **TRUE/FALSE** Write if the sentence is true (T) or false (F). If the sentence is false, change the <u>underlined</u> word or phrase to make it true. *(20 points: 4 points each)*

Example: <u>The United States</u> built the International Space Station. __F__ *Many countries*

1. The **lunar module** was named the *Eagle*. ____ _____

2. Scientists in the seventeenth century used **satellites** to look into space. ____ _____

3. Scientists send **space probes** to study other planets. ____ _____

4. **Satellites** orbit Earth. ____ _____

5. Rockets **cannot** reach outer space. ____ _____

B **MULTIPLE CHOICE** Choose the correct answer. *(20 points: 4 points each)*

Example: Scientists can live for months on a __b__.
    a. space probe   **b.** space station   **c.** space shuttle   **d.** telescope

6. Satellites can send ____ to and from Earth.
    a. booster rockets   **b.** space suits   **c.** launchpads   **d.** radio waves

7. The space shuttle carries or ____ scientists into space.
    a. collects   **b.** builds   **c.** records   **d.** transports

8. The word part *astro* in the word *astronaut* means ____.
    a. moon   **b.** sailor   **c.** star   **d.** station

9. Astronauts travel to the International Space Station on ____.
    a. the space shuttle   **c.** a space probe
    b. a command module   **d.** Sputnik

10. The astronauts on the moon ____.
    a. sent up rockets   **c.** took the space shuttle
    b. collected rocks   **d.** invented a telescope

C **COMPLETION** Complete the sentences. *(20 points: 4 points each)*

Example: The first space traveler was a _____*dog*_____ named Laika.

11. Neil Armstrong and Buzz Aldrin were the first people to land on _____.

12. Astronauts do experiments and make repairs while living on _____.

13. Ancient people recorded the movements of _____.

14. The astronauts wear _____ while traveling in space.

15. The USSR launched Sputnik, the first _____.

D **SHORT ANSWERS** Look at the time line. Answer the questions.
(20 points: 4 points each)

Events in Space Exploration

The USSR launches *Sputnik*.
1957

John Glenn is first American to orbit Earth.
1962

U.S. Space Shuttle makes first flight.
1981

1961
Yuri Gagarin is first person to orbit Earth.

1969
Apollo 11 astronauts walk on the moon.

1998
Building of International space station begins.

Example: When did the first American orbit Earth?

In 1962.

16. When did the USSR launch *Sputnik*?

_____

17. How long after the USSR launched *Sputnik* did a person orbit Earth?

_____

18. What important event happened in 1969?

_____

19. When did the space shuttle make its first flight?

_____

20. What happened seventeen years after the first space shuttle flight?

_____

E **WRITING** You are an astronaut on the International Space Station. Think about what you do in the morning, the afternoon, and at night. Tell about your day. What do you help scientists learn? Write a paragraph. *(20 points)*

_____

_____

_____

_____

_____

Name _____ Date _____

Student book pages 126–129

Grade

A TRUE/FALSE Write if the sentence is true (T) or false (F). If the sentence is false, change the underlined word or phrase to make it true. *(20 points: 4 points each)*

Example: **Volcanoes** carry away sediment. __F__ _____rivers_____

1. Sandstone is a type of **metamorphic** rock. ____ _____

2. Every **rock** has its own chemical formula. ____ _____

3. **Sedimentary rocks** are often found near rivers. ____ _____

4. Rocks are classified by how they **look**. ____ _____

5. **Weather** can affect rocks. ____ _____

B MULTIPLE CHOICE Choose the correct answer. *(20 points: 4 points each)*

Example: Igneous rocks form from __c__.
  a. fossils   **b.** sand   **c.** melted rocks   **d.** crystals

6. What is formed from bits of rock, shells, and fossils? ____
  a. igneous rocks   **c.** metamorphic rocks
  b. sedimentary rocks   **d.** diamonds

7. Rock breaks into tiny pieces because of ____.
  a. weathering   **b.** volcanoes   **c.** oceans   **d.** heat

8. Igneous rocks are often found near ____.
  a. rivers   **b.** fossils   **c.** shells   **d.** volcanoes

9. Heat and ____ can change one kind of rock into another kind of rock.
  a. dead plants   **b.** pressure   **c.** fossils   **d.** feldspar

10. A dead plant or animal can become ____.
  a. basalt   **c.** a fossil
  b. a metamorphic rock   **d.** a crystal

C COMPLETION Complete the sentences. *(20 points: 4 points each)*

Example: Basalt forms when _____lava_____ from a volcano cools quickly.

11. Diamond and feldspar are types of _____.

12. Limestone is a kind of _____ rock.

13. Heat and pressure can transform or _____ rock from one kind to another.

14. Rocks are classified into _____ groups.

15. Fossils are often found in _____ rocks.

D **SHORT ANSWERS** Look at the diagram. Answer the questions.
(20 points: 4 points each)

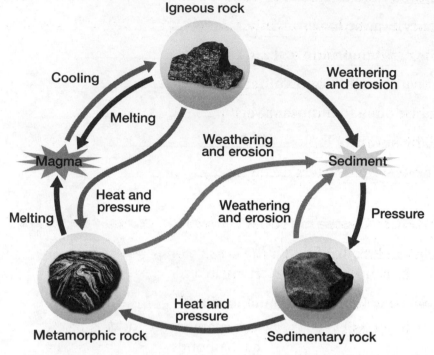

Igneous rock

Cooling

Melting

Weathering
and erosion

Weathering
and erosion

Magma

Sediment

Heat and
pressure

Weathering
and erosion

Pressure

Melting

Heat and
pressure

Metamorphic rock

Sedimentary rock

Example: What changes metamorphic rock into sediment?
_____ *weathering and erosion* _____

16. What changes other rocks into metamorphic rock?
_____

17. What happens to sediment that is under pressure?
_____

18. What two kinds of rock can melt and form magma?
_____

19. What forms when magma cools? _____

20. What kinds of rocks are affected by weathering?
_____

E **WRITING** You found a rock with the fossil of a plant in it. What kind of rock is it likely to be? Where might it be found? What happened between the time when the plant died and the time you found the rock? Write a paragraph. *(20 points)*

_____
_____
_____
_____
_____

GATEWAY TO SCIENCE Assessment Book • Copyright © Thomson Heinle

Name _____ Date _____

📖 Student book pages 130–133

A **TRUE/FALSE** Write if the sentence is true (T) or false (F). If the sentence is false, change the <u>underlined</u> word or phrase to make it true. *(20 points: 4 points each)*

Example: Earth's crust floats in the **lower mantle**. _F_ ___upper mantle___

1. Plate tectonics is the motion of Earth's **oceans**. ____ _____

2. The land on Earth was once one big **continent**. ____ _____

3. Seafloor spreading helps cause **continental drift**. ____ _____

4. A **plate** is part of Earth's crust. ____ _____

5. Earth's plates move **a few centimeters** each year. ____ _____

B **MULTIPLE CHOICE** Choose the correct answer. *(20 points: 4 points each)*

Example: The continents and the land under the ocean form Earth's __c__.
 a. core **b.** mantle **c.** crust **d.** layers

6. In seafloor spreading, continental plates move apart and form ____.
 a. a continent **b.** a fossil **c.** a crack **d.** a layer

7. Something that is *beneath* the ocean is ____ the ocean.
 a. close **b.** apart **c.** under **d.** against

8. The place where two of Earth's plates meet is ____.
 a. an oceanic ridge **c.** a continent
 b. a plate boundary **d.** the outer core

9. The word *tectonics* comes from a word that means ____.
 a. builder **b.** boundary **c.** drift **d.** rocks

10. Rocks in South America and Africa contain the same kind of ____.
 a. fossils **b.** magma **c.** plates **d.** borders

C **COMPLETION** Complete the sentences. *(20 points: 4 points each)*

Example: Earth's crust is made of pieces of solid _____rock_____.

11. Long ago, all the land on Earth was one big continent called _____.

12. When continental plates under the sea are moving apart, it is called

 _____.

13. Earth's _____ is harder and cooler than the lower mantle.

14. Earth's outer core is made of _____.

15. Fossils and rocks support the theory of _____.

D **SHORT ANSWERS** Look at the map. Answer the questions.
(20 points: 4 points each)

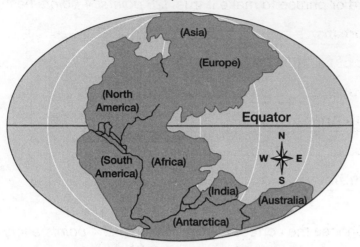

Long ago, all the land on Earth was one continent.

Example: Which continent is north of Europe?

     Asia

16. Which continent is west of Africa? _____

17. Is India north or south of the equator?

    _____

18. Which two whole continents lie north of the equator?

    _____

19. Which continents border South America?

    _____

20. Is India north or south of Antarctica?

    _____

E **WRITING** A scientist finds the same kind of fossils in Africa and in South America. What theory does this help explain? What does this evidence tell the scientist? Write a paragraph. *(20 points)*

_____

_____

_____

_____

_____

📖 Student book pages 134–137

Ⓐ TRUE/FALSE Write if the sentence is true (T) or false (F). If the sentence is false, change the underlined word or phrase to make it true. *(20 points: 4 points each)*

Example: **Fresh water** is in the oceans. __F__ _____salt water_____

1. A plain is a **flat** landform. ____ _____

2. Rivers flow **away from** the ocean. ____ _____

3. The largest continent is **North America**. ____ _____

4. A delta has a **square** shape. ____ _____

5. A **hill** is a type of landform. ____ _____

Ⓑ MULTIPLE CHOICE Choose the correct answer. *(20 points: 4 points each)*

Example: Mount Everest is in __c__.
   a. North America   **b.** Utah   **c.** Nepal   **d.** Egypt

6. Rivers contain _____.
   a. salt water   **b.** water vapor   **c.** fresh water   **d.** continents

7. Many mountains are caused by _____.
   a. rivers   **b.** volcanoes   **c.** deltas   **d.** water vapor

8. When a river reaches the ocean, the water _____.
   a. slows down   **c.** drops to the bottom
   b. builds up   **d.** flows underground

9. What contains a large part of our fresh water supply? _____
   a. canyons   **b.** oceans   **c.** deltas   **d.** groundwater

10. Water vapor is found in _____.
    a. rivers   **b.** canyons   **c.** the atmosphere   **d.** a delta

Ⓒ COMPLETION Complete the sentences. *(20 points: 4 points each)*

Example: The longest river on Earth is the _____Nile_____.

11. The largest canyon on Earth is in _____.

12. Wind and moving _____ can form a canyon.

13. A _____ is a mass of ice that moves across Earth's surface.

14. Sand and gravel accumulate or _____ to form a delta.

15. The topography of Earth is the _____ of its surface.

Name _____ Date _____

D SHORT ANSWERS Look at the maps. Answer the questions.
(20 points: 4 points each)

A Topographic Map of Sailboat Island

A Picture of Sailboat Island

Example: Which is the lowest point on Sailboat Island?

_____ *Sunset Beach* _____

16. How high is Palm Tree Point?

_____

17. How high is Coconut Hill?

_____

18. There is a low spot between the two hills. What is its elevation?

_____

19. How much higher is Palm Tree Point than Coconut Hill?

_____

20. What is the elevation of Sunset Beach?

_____

E WRITING You just visited Egypt. You saw the delta where the Nile River meets the Mediterranean Sea. Write a letter to a friend. Tell what caused this landform. Write a paragraph. *(20 points)*

_____

_____

_____

_____

_____

Name _____ Date _____

📖 Student book pages 138–141

⬚ Grade

A **TRUE/FALSE** Write if the sentence is true (T) or false (F). If the sentence is false, change the underlined word or phrase to make it true. *(20 points: 4 points each)*

Example: **A vent** is a shaking of Earth's crust. __F__ __an earthquake__

1. Many earthquakes happen at **plate boundaries**. ____ _____

2. **A volcano** can form new land. ____ _____

3. The layer of rock on the Earth's surface is **the mantle**. ____ _____

4. Earth's crust is divided into **plates**. ____ _____

5. Volcanoes form **in the center** of Earth's plates. ____ _____

B **MULTIPLE CHOICE** Choose the correct answer. *(20 points: 4 points each)*

Example: Most earthquakes happen along __b__.
   a. craters   b. faults   c. vents   d. waves

6. The hot material that flows down the sides of a volcano is ____.
   a. lava   b. crust   c. a tsunami   d. a plate

7. The place inside Earth where an earthquake starts is the ____.
   a. focus   b. epicenter   c. lava   d. vent

8. The word part *nami* in the word *tsunami* means ____.
   a. crater   b. lava   c. wave   d. eruption

9. Volcanoes can form at hot spots in Earth's ____.
   a. epicenter   b. waves   c. focus   d. mantle

10. The bowl-shaped hole in the top of a volcano is its ____.
   a. fault   b. crater   c. magma   d. focus

C **COMPLETION** Complete the sentences. *(20 points: 4 points each)*

Example: A tsunami is a huge _____wave_____.

11. When Earth's plates converge, they move _____.

12. The place above an earthquake's focus is the _____.

13. The Hawaiian Islands are on one of Earth's _____.

14. The word _____ means "a crack or split in Earth's surface."

15. Earthquakes happen when _____ builds up at plate boundaries.

D **SHORT ANSWERS** Look at the drawings. Answer the questions.
(20 points: 4 points each)

A

Plate
Magma
Plate

B

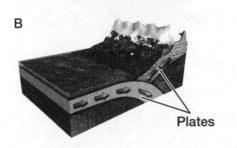

Plates

C

Plates

Example: In what direction are the plates in drawing A moving?
_____*apart*_____

16. What kind of boundary does drawing B show?
_____

17. Are the plates in drawing C moving over each other or past each other?
_____

18. What shows which way the plates are moving in these drawings?
_____

19. Which drawing shows plates diverging?
_____

20. Which drawing shows magma rising to the surface?
_____

E **WRITING** You are visiting the Hawaiian Islands. You see volcanoes. What causes a volcano? Why are many volcanoes there? Write an e-mail to a friend. Explain what you saw. Write a paragraph. *(20 points)*

_____

_____

_____

_____

_____

Name _____ Date _____

📖 Student book pages 142–145

Grade

A **TRUE/FALSE** Write if the sentence is true (T) or false (F). If the sentence is false, change the <u>underlined</u> word or phrase to make it true. *(20 points: 4 points each)*

Example: Wind can cause **chemical weathering**. ___F___ ___mechanical weathering___

1. Water can move **sediment** to new places. ____ _____

2. **Chemical weathering** can break down rocks. ____ _____

3. **Deposition** builds new landforms. ____ _____

4. Weathering changes Earth's surface **quickly**. ____ _____

5. **Wind** can build up sand dunes. ____ _____

B **MULTIPLE CHOICE** Choose the correct answer. *(20 points: 4 points each)*

Example: A glacier is a slowly moving mass of __c__.
    a. rock   **b.** wind   **c.** ice   **d.** sediment

6. In deposition, ____ builds up in a place.
    a. ice   **b.** water   **c.** carbon dioxide   **d.** sediment

7. Oxygen and carbon dioxide can cause ____.
    a. mechanical weathering   **c.** glaciers
    b. chemical weathering   **d.** deposition

8. The Grand Canyon was formed by ____.
    a. deposition   **c.** chemical weathering
    b. erosion   **d.** glaciers

9. As a glacier moves, ____ inside it can wear down a mountain.
    a. plants   **b.** cracks   **c.** chemicals   **d.** sand and rocks

10. Tiny pieces of rock wear down bigger rocks through ____.
    a. abrasion   **c.** ice wedging
    b. deposition   **d.** chemical weathering

C **COMPLETION** Complete the sentences. *(20 points: 4 points each)*

Example: A plant's _____roots_____ can help crack a rock.

11. Ocean _____ erode sand from beaches.

12. Ice wedging happens when water freezes in a _____ in a rock.

13. A mountain's shape can tell us where a _____ formed in the past.

14. _____ carries dust and soil through the air.

15. Earthquakes and _____ can change Earth's surface quickly.

Name _____   Date _____

D **SHORT ANSWERS** Look at the photos. Answer the questions.
(20 points: 4 points each)

A

B

C

Example: What caused the landform in photo A?
_____ *wind* _____

16. What landform is the effect of deposition in photo A?
_____

17. What is causing the rock to crack in photo C?
_____

18. What process caused the canyon in photo B to form?
_____

19. The river in photo B carries chemicals. They can break down a rock. What is this
effect called? _____

20. The river in photo B carries tiny bits of rock. The bits cause rock to rub off the canyon
walls. What is this effect called? _____

E **WRITING** You build a stone house in a windy, rainy place. What will happen to
the house over time? What will the wind do? What will the rain do? Write a paragraph. *(20 points)*

_____

_____

_____

_____

_____

GATEWAY TO SCIENCE Assessment Book • Copyright © Thomson Heinle

Name _____   Date _____

Grade ⬜

A TRUE/FALSE Write if the sentence is true (T) or false (F). If the sentence is false, change the <u>underlined</u> word or phrase to make it true. *(20 points: 4 points each)*

Example: The stratosphere is **below** the troposphere. __F__ _____above_____

1. Animals and plants need **gases** to live. ____ _____

2. **Large** amounts of trace gases are in air. ____ _____

3. Earth's **atmosphere** makes life possible. ____ _____

4. The atmosphere is **inside** Earth. ____ _____

5. The atmosphere is composed of **one type** of gas. ____ _____

B MULTIPLE CHOICE Choose the correct answer. *(20 points: 4 points each)*

Example: Plants need __d__.
   a. pollution   **b.** fog   **c.** wind   **d.** carbon dioxide

6. The troposphere is ____.
   a. warm   **b.** cold   **c.** common   **d.** very thin

7. Animals make ____ as waste.
   a. water vapor   **b.** oxygen   **c.** carbon dioxide **d.** fog

8. Plants can live only in the ____.
   a. stratosphere **b.** troposphere   **c.** exosphere   **d.** mesosphere

9. The word part *sphere* in the word *atmosphere* means ____.
   a. thin   **b.** trace   **c.** round   **d.** cycle

10. Plants make ____ as waste.
   a. fog   **b.** nitrogen   **c.** oxygen   **d.** carbon dioxide

C COMPLETION Complete the sentences. *(20 points: 4 points each)*

Example: Two places plants get carbon dioxide are __air and water__.

11. The most common gas in the air is _____.

12. The word part *atmos* in *atmosphere* means _____.

13. Animals get _____ from the air or from water.

14. Gases that are not common in the air are called _____.

15. The lowest layer of the atmosphere is the _____.

D **SHORT ANSWERS** Look at the pie chart. Answer the questions.
(20 points: 4 points each)

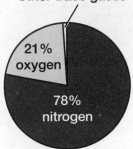

1% carbon dioxide,
water vapor, and
other trace gases

21% oxygen

78% nitrogen

Gases in the Atmosphere

Example: Does the atmosphere have less oxygen or less carbon dioxide?

_____ *carbon dioxide* _____

16. What percentage of the air is nitrogen?

_____

17. What percentage of the air is oxygen?

_____

18. What percentage of the air is trace gases?

_____

19. What are two trace gases in the atmosphere?

_____

20. Which section of the pie chart is the smallest?

_____

E **WRITING** Name the layer of the atmosphere where you live. Tell why you live in
that layer. How does it feel? How is it different from other layers of the atmosphere?
Write a paragraph. *(20 points)*

_____

_____

_____

_____

_____

Name _____ Date _____

📖 Student book pages 150–153

A **TRUE/FALSE** Write if the sentence is true (T) or false (F). If the sentence is false, change the <u>underlined</u> word or phrase to make it true. *(20 points: 4 points each)*

Example: The United States has **one** climate. __F__ __more than one__

1. **Low-pressure** systems often bring storms. ____ _____

2. **Gravity** acts on large water droplets in the air. ____ _____

3. Alaska has a **tropical** climate. ____ _____

4. Hail is a type of **precipitation**. ____ _____

5. Climate changes very **quickly**. ____ _____

B **MULTIPLE CHOICE** Choose the correct answer. *(20 points: 4 points each)*

Example: Atmospheric pressure is the __c__ of the air in a certain place.
   a. climate   **b.** water vapor   **c.** weight   **d.** temperature

6. Some weather maps show or ____ fronts.
   a. move   **b.** display   **c.** develop   **d.** expand

7. High-pressure masses usually ____ low-pressure masses.
   a. expand   **b.** block   **c.** go around   **d.** collect

8. Low pressure areas must ____ high pressure areas.
   a. connect   **b.** collect   **c.** go around   **d.** build up

9. Droplets of water in the air collect around particles of ____.
   a. snow and ice   **b.** wind   **c.** water vapor   **d.** dust and salt

10. Isobars connect areas with the same amount of ____.
   a. hail   **b.** pressure   **c.** water vapor   **d.** gravity

C **COMPLETION** Complete the sentences. *(20 points: 4 points each)*

Example: A warm front is a warm air mass that is pushing into a ____cold air mass____.

11. Rain and snow are types of _____.

12. Climate is an area's _____ weather over the years.

13. Millions of particles of dust and water vapor come together. Together, they form a

_____.

14. The word part *bar* in the word *isobar* means _____.

15. The average amount of precipitation is part of an area's _____.

Name _____   Date _____

D **SHORT ANSWERS** Look at the weather map. Answer the questions.
(20 points: 4 points each)

A Weather Map

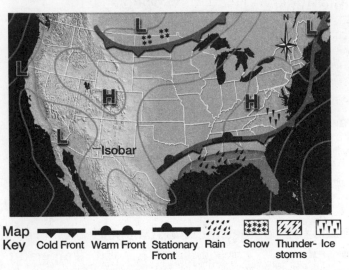

Example: Find the cold front near the top of the map. What kind of weather will be

north of the front? _____ snowy _____

16. What kind of pressure is north of the cold front at the top of the map?

_____

17. Find the stationary front near the bottom of the map. What kind of weather will be

south of the front? _____

18. Find the front at the top center of the map. What direction is it moving?

_____

19. Look along the west side of the map. Is that area under high or low pressure?

_____

20. What kinds of fronts are shown on the map?

_____

E **WRITING** Explain the difference between weather and climate. Which one
changes quickly? Which one changes over time? Write a paragraph. *(20 points)*

_____

_____

_____

_____

GATEWAY TO SCIENCE Assessment Book • Copyright © Thomson Heinle

Name _____ Date _____

📖 Student book pages 154–157

A **TRUE/FALSE** Write if the sentence is true (T) or false (F). If the sentence is false, change the <u>underlined</u> word to make it true. *(20 points: 4 points each)*

Example: A tropical disturbance is a storm over <u>**land**</u>. __F__ _____water_____

1. A <u>**watch**</u> means severe weather has been seen. ____ _____

2. Tornadoes have strong **updrafts**. ____ _____

3. A hurricane rotates around its **eye**. ____ _____

4. <u>**Cyclones**</u> form in the North Atlantic Ocean or the East Pacific Ocean. ____

   _____

5. A <u>**tornado**</u> is a spinning column of moist air that touches the ground. ____

   _____

B **MULTIPLE CHOICE** Choose the correct answer. *(20 points: 4 points each)*

Example: A __a__ forms in the North Pacific Ocean or the South China Sea.
   a. typhoon   **b.** tornado   **c.** flood   **d.** warning

6. Moist air holds a lot of ____.
   a. lightning   **b.** water vapor   **c.** land   **d.** wind speed

7. When a tornado forms over water, it is called a ____.
   a. hurricane   **b.** cyclone   **c.** updraft   **d.** waterspout

8. A ____ hangs below a main cloud.
   a. tropical storm   **b.** siren   **c.** wall cloud   **d.** downdraft

9. A tornado watch means or ____ that severe weather is possible.
   a. indicates   **b.** forms   **c.** touches   **d.** rotates

10. The winds in a ____ are higher than in a tropical disturbance.
   a. wall cloud   **c.** flood
   b. updraft   **d.** tropical depression

C **COMPLETION** Complete the sentences. *(20 points: 4 points each)*

Example: A storm surge is water rushing _____over land_____.

11. Hurricanes, cyclones, and typhoons are types of _____ storms.

12. When a tropical depression becomes a tropical storm, scientists give it

   a _____.

13. When warm, moist air cools, the water vapor turns to _____.

14. Watches and _____ are given for severe weather.

15. Thunderstorms usually form and go away _____.

Name _____ Date _____

D **SHORT ANSWERS** Look at the diagram. Answer the questions.
(20 points: 4 points each)

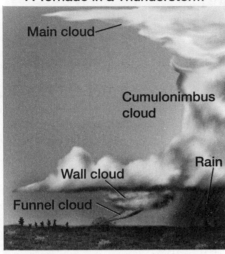

A Tornado in a Thunderstorm

Main cloud

Cumulonimbus cloud

Rain

Wall cloud

Funnel cloud

Example: Is the funnel cloud above or below the main cloud?
_____*below*_____

16. In what kind of cloud does a thunderstorm form?
_____

17. What kind of cloud hangs down from the main cloud?
_____

18. Is the wall cloud above or below the funnel cloud?
_____

19. What kind of precipitation falls from these clouds?
_____

20. What do we call a funnel cloud when it touches ground?
_____

E **WRITING** Imagine that you are a TV weather reporter. There was a bad thunderstorm in your city. Tell everyone how the thunderstorm formed. Tell about air temperature. Tell about updrafts and downdrafts. Write a paragraph. *(20 points)*

_____

_____

_____

_____

_____

Grade

📖 Student book pages 158–161

A **TRUE/FALSE** Write if the sentence is true (T) or false (F). If the sentence is false, change the <u>underlined</u> word or phrase to make it true. *(20 points: 4 points each)*

Example: Fossil fuels form from **minerals**. __F__ _dead plants and animals_

1. Wind energy can produce **electricity**. ____ _____

2. Biomass is a **nonrenewable resource**. ____ _____

3. **Fossil fuels** are natural resources. ____ _____

4. **Plants** are natural resources. ____ _____

5. Gold and copper are **fossil fuels**. ____ _____

B **MULTIPLE CHOICE** Choose the correct answer. *(20 points: 4 points each)*

Example: Geysers can produce __b__.
    a. water energy   **c.** atomic energy
    b. geothermal energy   **d.** solar energy

6. Solar energy comes from ____.
    a. wind   **b.** water   **c.** sunlight   **d.** biomass

7. Coal, petroleum, and natural gas are ____.
    a. minerals   **c.** renewable resources
    b. wind machines   **d.** fossil fuels

8. The word part *geo* in the word *geothermal* means ____.
    a. heat   **b.** Earth   **c.** energy   **d.** sunlight

9. Atomic energy plants make ____.
    a. solar energy   **b.** biomass   **c.** nuclear energy   **d.** minerals

10. Fossil fuels are ____.
    a. nonrenewable resources   **c.** renewable resources
    b. part of biomass   **d.** made in atomic energy plants

C **COMPLETION** Complete the sentences. *(20 points: 4 points each)*

Example: People consume or ____use____ natural resources.

11. It takes millions of years to form _____ fuels.

12. A dam traps the power of moving _____.

13. Wind turns the blades on a _____.

14. When we reduce, we use fewer _____.

15. Biomass is used to make _____.

Name _____ Date _____

D **SHORT ANSWERS** Look at the pie chart. Answer the questions.
(20 points: 4 points each)

Energy Use in Developed Countries

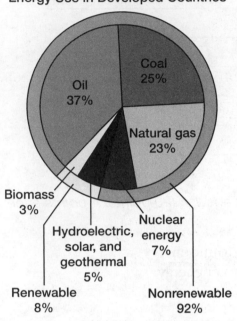

Example: How much of the energy used in developed countries comes from oil?

_____37%_____

16. How much of the energy comes from renewable resources?

_____

17. How much more energy comes from oil than from biomass?

_____

18. What percentage of energy comes from nonrenewable resources?

_____

19. How much less energy comes from renewable resources than from coal?

_____

20. What percent of energy is provided by fossil fuels (coal, oil, and natural gas)?

_____

E **WRITING** Name three kinds of fossil fuels. Tell why they are nonrenewable. Write a paragraph. *(20 points)*

_____

_____

_____

_____

_____

GATEWAY TO SCIENCE Assessment Book • Copyright © Thomson Heinle

Name _____ Date _____

📖 Student book pages 102–161

Grade ▢

Ⓐ TRUE/FALSE Write if the sentence is true (T) or false (F). If the sentence is false, change the underlined word or phrase to make it true. *(20 points: 2 points each)*

Example: Space is the area beyond **the sun**. _F_ _____Earth_____

1. There are **thousands** of galaxies in the universe. ____ _____

2. A **dwarf planet** has a glowing tail. ____ _____

3. The Bay of Fundy has a **large** tidal range. ____ _____

4. Sandstone is a type of **igneous** rock. ____ _____

5. A plain is a **flat** landform. ____ _____

6. Water can move **sediment** to new places. ____ _____

7. **Low-pressure** systems often can bring storms. ____ _____

8. Water energy can produce **electricity**. ____ _____

9. Every **rock** has its own chemical formula. ____ _____

10. **Mechanical weathering** can break down rocks. ____ _____

Ⓑ MULTIPLE CHOICE Choose the correct answer. *(40 points: 2 points each)*

Example: Galaxies are the _c_ objects in space.
    a. smallest   **b.** nearest   **c.** largest   **d.** emptiest

11. The Milky Way is a ____.
    a. comet   **b.** planet   **c.** galaxy   **d.** telescope

12. A main sequence star burns ____ as fuel.
    a. oxygen   **b.** neutrons   **c.** hydrogen   **d.** gravity

13. Jupiter's Great Red Spot is a giant ____.
    a. comet   **b.** storm   **c.** asteroid   **d.** moon

14. The warmest season on Earth is ____.
    a. spring   **b.** fall   **c.** summer   **d.** winter

15. A ____ occurs when the moon passes between the sun and Earth.
    a. lunar eclipse   **b.** solar eclipse   **c.** neap tide   **d.** tidal bulge

B **MULTIPLE CHOICE,** continued

16. Satellites can send ____ to and from Earth.

    a. booster rockets **b.** space probes **c.** launchpads **d.** radio waves

17. ____ form from bits of rock, shells, and fossils.

    a. Igneous rocks **c.** Metamorphic rocks
    b. Sedimentary rocks **d.** Diamonds

18. In seafloor spreading, continental plates move apart and form ____.

    a. a continent **b.** a fossil **c.** a crack **d.** a tsunami

19. Rivers contain ____.

    a. salt water **b.** water vapor **c.** fresh water **d.** erosion

20. The hot material that flows down the sides of a volcano is ____.

    a. lava **b.** crust **c.** a tsunami **d.** a moraine

21. In deposition, ____ builds up in a place.

    a. glaciers **b.** water **c.** carbon dioxide **d.** sediment

22. The troposphere is ____.

    a. warm **b.** cold **c.** heavy **d.** very thin

23. Some weather maps show or ____ fronts.

    a. move **b.** display **c.** develop **d.** color

24. Moist air holds a lot of ____.

    a. lightning **b.** water vapor **c.** hail **d.** wind speed

25. Solar energy comes from ____.

    a. wind **b.** water **c.** sunlight **d.** oil

26. Some telescopes collect ____ with lenses.

    a. light waves **b.** objects **c.** light-years **d.** gases

27. A comet's glowing tail is made of ____.

    a. metal **b.** minerals **c.** gas **d.** ice

28. The Space Shuttle carries or ____ scientists into space.

    a. collects **b.** builds **c.** probes **d.** transports

Name _____ Date _____

B MULTIPLE CHOICE, continued

29. The place inside Earth where an earthquake starts is the ____.
   a. focus   b. epicenter   c. lava   d. magma

30. High-pressure masses usually ____ low-pressure masses.
   a. protect   b. block   c. go around   d. collect

C COMPLETION Complete the sentences. *(20 points: 2 points each)*

Example: Nothing can escape from a _____black hole_____.

31. A white dwarf shrinks, leaving behind a new _____.

32. The lengths of night and day change with the _____.

33. Neil Armstrong and Buzz Aldrin were the first Americans to land on

   _____.

34. Long ago, all of Earth's land was one big continent called _____.

35. When Earth's plates converge, they move _____.

36. The most common gas in the air is _____.

37. Hurricanes, cyclones, and typhoons are all _____ storms.

38. Continental plates under the ocean moving apart is called _____.

39. The word part *atmos* in *atmosphere* means _____.

40. Part of the sun's light is blocked in a _____ eclipse.

Name _____ Date _____

D SHORT ANSWERS Look at the maps. Answer the questions.
(20 points: 2 points each)

A Topographic Map of Sailboat Island

A Picture of Sailboat Island

Example: Which is the lowest point on Sailboat Island?

    Sunset Beach

41. How high is Palm Tree Point?

    _____

42. How high is Coconut Hill?

    _____

43. There is a low place between the two hills. What is its elevation?

    _____

44. How much higher is Palm Tree Point than Coconut Hill?

    _____

45. What is the elevation of Sunset Beach?

    _____

GATEWAY TO SCIENCE Assessment Book • Copyright © Thomson Heinle

Name _____   Date _____

D SHORT ANSWERS, continued

Look at the time line. Answer the questions.

Events in Space Exploration

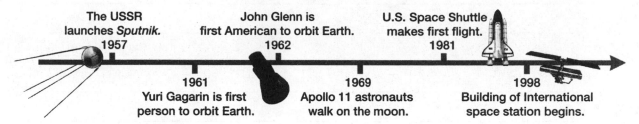

The USSR launches *Sputnik.* 1957

John Glenn is first American to orbit Earth. 1962

U.S. Space Shuttle makes first flight. 1981

1961 Yuri Gagarin is first person to orbit Earth.

1969 Apollo 11 astronauts walk on the moon.

1998 Building of International space station begins.

46. In what year did the USSR launch *Sputnik?*

_____

47. How long after *Sputnik* launched did a person orbit Earth?

_____

48. What event happened in 1969?

_____

49. When did the space shuttle first fly?

_____

50. What happened 17 years after the first space shuttle flight?

_____

E **WRITING ASSESSMENT** Write paragraphs. *(100 points: 25 points each)*

51. It is a summer day in North America. Describe the tilt of Earth's axis. What is the weather like? Are the days long or short?

_____

_____

_____

_____

_____

52. Describe the processes that sediment will go through to become igneous rock.

_____

_____

_____

_____

_____

53. How can ice change Earth's surface? What can ice do to a rock? What can ice do to a mountain?

_____

_____

_____

_____

_____

54. You see a tropical disturbance on a TV weather map. Name the stages the storm will go through as it becomes a hurricane. Describe the stages.

_____

_____

_____

_____

_____

GATEWAY TO SCIENCE Assessment Book • Copyright © Thomson Heinle

A TRUE/FALSE Write if the sentence is true (T) or false (F). If the sentence is false, change the <u>underlined</u> word to make it true. *(20 points: 4 points each)*

Example: Taste is a **chemical property** of matter. __F__ __physical property__

1. Matter has **four** states. ____ _____

2. A liquid has a definite **shape**. ____ _____

3. A **property** is a characteristic or trait. ____ _____

4. Water vapor is a **liquid**. ____ _____

5. A **gas** has a definite shape. ____ _____

B MULTIPLE CHOICE Choose the correct answer. *(20 points: 4 points each)*

Example: You can see the __a__ of an object.
   a. color   **b.** odor   **c.** smell   **d.** taste

6. The particles in ____ are very far away from one another.
   a. a solid   **b.** a liquid   **c.** a gas   **d.** rust

7. Your sense of smell can tell you the ____ of an object.
   a. shape   **b.** taste   **c.** odor   **d.** texture

8. Rusting is an example of ____.
   a. physical change   **b.** chemical change   **c.** texture   **d.** a sense

9. You pour a liquid from a bottle into a glass. Its ____ changes.
   a. volume   **b.** odor   **c.** texture   **d.** shape

10. Your sense of touch can help you feel the ____ of an object.
   a. color   **b.** taste   **c.** texture   **d.** state

C COMPLETION Complete the sentences. *(20 points: 4 points each)*

Example: Burning is an example of a _____chemical_____ change.

11. You can see the color of an object through your sense of _____.

12. Shape, size, and texture are _____ properties of matter.

13. You can recognize the _____ of matter by observing the matter.

14. You can observe the _____ of matter during a chemical change.

15. Smell, hearing, and touch are examples of _____.

Name _____ Date _____

D **SHORT ANSWERS** Look at the pictures. Answer the questions.
(20 points: 4 points each)

Solid
A

Liquid
B

Gas
C

D

Example: Which picture shows a gas? _____ *picture C* _____

16. What kind of change does picture D show?

_____

17. In what state is the matter in picture A?

_____

18. In which picture are the particles closer together, picture A or picture B?

_____

19. Which states would change shape if you change their containers?

_____

20. For which state does the container have a lid?

_____

E **WRITING** Think of an apple. Describe its properties. Tell which senses you would use to observe its properties. Write a paragraph. *(20 points)*

_____

_____

_____

_____

_____

Name _____ Date _____

Student book pages 166–169

Grade

A TRUE/FALSE Write if the sentence is true (T) or false (F). If the sentence is false, change the <u>underlined</u> word or phrase to make it true. *(20 points: 4 points each)*

Example: We express **volume** in grams per cubic centimeter. __F__ ____density____

1. The **melting point** of ice is 0°C. ____ _____

2. Different materials have different masses for the same **weight**. ____ _____

3. If you heat liquid water, it **melts**. ____ _____

4. All objects have **mass**. ____ _____

5. Some **solids** can float in liquids. ____ _____

B MULTIPLE CHOICE Choose the correct answer. *(20 points: 4 points each)*

Example: We measure __a__ with a graduated cylinder.
   a. volume   **b.** mass   **c.** temperature   **d.** density

6. When you put a solid in a liquid, ____ pushes up on the object.
   a. density   **b.** force of gravity   **c.** buoyant force   **d.** volume

7. The amount of space something occupies is its ____.
   a. boiling point   **b.** volume   **c.** mass   **d.** density

8. Steam is a ____.
   a. solid   **b.** liquid   **c.** gas   **d.** boiling point

9. We measure mass with a ____.
   a. thermometer   **c.** liquid
   b. balance   **d.** graduated cylinder

10. Ice ____ when it changes to water.
   a. melts   **b.** floats   **c.** boils   **d.** sinks

C COMPLETION Complete the sentences. *(20 points: 4 points each)*

Example: We measure boiling points with a ___thermometer___.

11. The _____ of water is 100°C.

12. Mass divided by volume equals _____.

13. The force of gravity pushes _____ on a floating object.

14. The force on an object caused by gravity is its _____.

15. An object _____ when the buoyant force is greater than the force of gravity.

D **SHORT ANSWERS** Look at the table. Answer the questions.
(20 points: 4 points each)

Melting Points and Boiling Points

Substance	Melting point	Boiling point
water	0°C	100°C
salt	801°C	1,413°C
oxygen	-218°C	-189°C

Example: Is the boiling point of salt higher or lower than the boiling point of water?

_____ higher _____

16. Which substance has the lowest melting point?

_____

17. Which substance has the highest boiling point?

_____

18. How much higher is the boiling point of salt than the boiling point of water?

_____

19. Which substance has the lowest boiling point?

_____

20. How many degrees higher is salt's boiling point than its melting point?

_____

E **WRITING** You take a trip to the moon. Your weight changed. What happened? How did it change? Explain why it changed. Did your mass change, too? Explain why. Write a paragraph. *(20 points)*

_____

_____

_____

_____

_____

Name _____   Date _____

📖 Student book pages 170–173

Grade

A **TRUE/FALSE** Write if the sentence is true (T) or false (F). If the sentence is false, change the underlined word to make it true. *(20 points: 4 points each)*

Example: Electrons have a **positive** charge. __F__ ____negative____

1. Protons are **smaller** than atoms. ____ _____

2. Water forms when **molecules** join together. ____ _____

3. Germanium is named after **Germany**. ____ _____

4. All matter is made up of **minerals**. ____ _____

5. Gold is a **nonmetal**. ____ _____

B **MULTIPLE CHOICE** Choose the correct answer. *(20 points: 4 points each)*

Example: Oxygen is a __d__.
   a. metal   **b.** proton   **c.** nucleus   **d.** nonmetal

6. The particles that circle an atom's nucleus are ____.
   a. protons   **b.** electrons   **c.** neutrons   **d.** molecules

7. Each ____ has a box on the periodic table.
   a. molecule   **b.** proton   **c.** element   **d.** nucleus

8. At room temperature, hydrogen is a ____.
   a. molecule   **b.** liquid   **c.** gas   **d.** metal

9. Every element has its own ____ in the periodic table.
   a. symbol   **b.** color   **c.** charge   **d.** molecule

10. An element is made of one kind of ____.
   a. nucleus   **b.** molecule   **c.** metal   **d.** atom

C **COMPLETION** Complete the sentences. *(20 points: 4 points each)*

Example: The number of protons in the nucleus is an atom's ____atomic number____.

11. A proton has a _____ electrical charge.

12. At room temperature, water is a _____.

13. When two atoms join together or _____ they form a molecule.

14. Two groups on the periodic table are metals and _____.

15. Everything you can see, smell, taste, and touch is _____.

D **SHORT ANSWERS** Look at the diagram. Answer the questions.
(20 points: 4 points each)

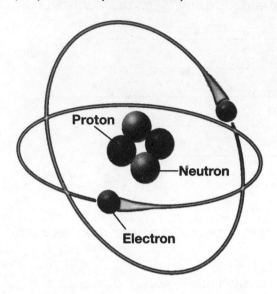

Example: What structure does this diagram show?
_____ *an atom* _____

16. What are protons, neutrons, and electrons?
_____

17. What particles are in the center of the atom?
_____

18. What particles circle the center of the atom?
_____

19. How many protons does this atom have?
_____

20. How many neutrons does this atom have?

_____

E **WRITING** Think about becoming very, very small. You take a trip through an atom. What do you see? Name the parts of the atom. Tell where you find them. What are their charges? Write a paragraph. *(20 points)*

_____
_____
_____
_____
_____

Name _____ Date _____

Student book pages 174–177

Grade

A **TRUE/FALSE** Write if the sentence is true (T) or false (F). If the sentence is false, change the underlined word or phrase to make it true. *(20 points: 4 points each)*

Example: Aluminum is **a heterogeneous mixture**. __F__ ___an element___

1. During a **physical change**, new substances form. ____ _____

2. Elements in a **compound** are joined chemically. ____ _____

3. A compound is a **pure substance**. ____ _____

4. Sugar dissolves in **tea**. ____ _____

5. **A mixture** is a pure substance. ____ _____

B **MULTIPLE CHOICE** Choose the correct answer. *(20 points: 4 points each)*

Example: Carbon and oxygen combine to form ___c___.
   a. salt   **b.** sugar   **c.** carbon dioxide   **d.** air

6. When wood burns, a ____ happens.
   a. physical change   **b.** pure substance   **c.** solution   **d.** chemical change

7. A pure substance is made of one kind of ____.
   a. solution   **b.** mixture   **c.** compound   **d.** particle

8. Some pure substances are ____.
   a. mixtures   **b.** compounds   **c.** suspensions   **d.** solutions

9. A suspension is a kind of ____.
   a. homogeneous mixture   **c.** compound
   b. heterogeneous mixture   **d.** element

10. The substances in ____ are not joined chemically.
   a. a compound   **b.** a mixture   **c.** an element   **d.** a fire

C **COMPLETION** Complete the sentences. *(20 points: 4 points each)*

Example: Salad dressing is an example of a ___heterogeneous___ mixture.

11. During a _____ change, the identity of the substance does not change.

12. In a _____ one substance dissolves into another substance.

13. When ice melts, a _____ change happens.

14. The word part *hetero* in the word *heterogeneous* means _____.

15. Sodium and chlorine join to form the compound _____.

D **SHORT ANSWERS** Look at the table. Answer the questions.
(20 points: 4 points each)

Compound	Elements Joined in the Compound
water	hydrogen and oxygen
table salt (sodium chloride)	sodium and chlorine
sugar	carbon, hydrogen, and oxygen
baking soda	sodium, hydrogen, carbon, and oxygen
chalk	calcium, carbon, and oxygen

Example: Which compounds contain hydrogen?
_____water, sugar, and baking soda_____

16. Which compounds are formed from the fewest elements?
_____

17. Which compounds contain oxygen and carbon?
_____

18. Which compound contains more elements: sugar or baking soda?
_____

19. Which compound contains chlorine?
_____

20. What elements do sugar and water have in common?
_____

E **WRITING** You stir some sugar into some water. Describe what happens. What kind of substances do you start with? What are they made of? What kind of substance do you end up with? Is this a physical change or a chemical change? Write a paragraph. *(20 points)*

_____

_____

_____

_____

_____

A TRUE/FALSE Write if the sentence is true (T) or false (F). If the sentence is false, change the <u>underlined</u> word or phrase to make it true. *(20 points: 4 points each)*

Example: A compound contains **only one kind** of atom. __F__ *more than one kind*

1. An **endothermic** reaction gives off energy. ____ _____

2. Sodium is an **element**. ____ _____

3. Atoms can join together to form **molecules**. ____ _____

4. All matter is made of **atoms**. ____ _____

5. Matter **can be** created. ____ _____

B MULTIPLE CHOICE Choose the correct answer. *(20 points: 4 points each)*

Example: Water is __c__ of hydrogen and oxygen.
   a. an element   **b.** an atom   **c.** a compound   **d.** a symbol

6. Every compound has ____.
   a. a symbol   **b.** a formula   **c.** an equal sign   **d.** an equation

7. Iron and oxygen combine to form ____.
   a. sodium   **b.** water   **c.** chlorine   **d.** rust

8. Atoms and molecules come together or break apart in ____.
   a. chemical reactions   **c.** symbols
   b. chemical equations   **d.** mathematical equations

9. When something burns, it combines with ____.
   a. sodium   **b.** chlorine   **c.** water   **d.** oxygen

10. When atoms form molecules, they can gain or lose ____.
   a. a nucleus   **b.** symbols   **c.** electrons   **d.** mass

C COMPLETION Complete the sentences. *(20 points: 4 points each)*

Example: The law of __conservation of mass__ says that matter cannot be destroyed.

11. Chemical _____ break the chemical bonds between atoms.

12. The substance on the right side of a chemical formula is called the _____.

13. The _____ for oxygen is O.

14. A molecule is formed from two or more _____.

15. A math equation states that two amounts are _____.

D **SHORT ANSWERS** Look at the chemical equation. Answer the questions.
(20 points: 4 points each)

$$2H_2 + O_2 \rightarrow 2H_2O$$

Example: How many hydrogen atoms are in the reactants?

_____4_____

16. How many hydrogen atoms are in the product?

_____

17. How many molecules are in the reactants?

_____

18. How many molecules are in the product?

_____

19. What does the small number *2* after a symbol tell you?

_____

20. What does the yield sign tell you?

_____

E **WRITING** The symbol for sodium is Na. The symbol for chlorine is Cl. The chemical equation for table salt is $2Na + Cl_2 \rightarrow 2NaCl$. Explain what happens in the
chemical reaction that produces salt. Write a paragraph. *(20 points)*

_____

_____

_____

_____

_____

GATEWAY TO SCIENCE Assessment Book • Copyright © Thomson Heinle

📖 Student book pages 182–185

Ⓐ TRUE/FALSE Write if the sentence is true (T) or false (F). If the sentence is false, change the <u>underlined</u> word to make it true. *(20 points: 4 points each)*

Example: A beta particle is usually **a proton**. __F__ ____an electron____

1. Radiation can help treat **cancer**. ____ _____

2. Radioactive materials give off **uranium** and particles. ____ _____

3. Uranium gives off **energy**. ____ _____

4. Exposure to **radiation** can be dangerous. ____ _____

5. Antoine-Henri Becquerel discovered **uranium**. ____ _____

Ⓑ MULTIPLE CHOICE Choose the correct answer. *(20 points: 4 points each)*

Example: Gamma rays are a type of __d__.
   a. X-ray   **b.** proton   **c.** particle   **d.** energy

6. Doctors use ____ to check for broken bones.
   a. X-rays   **b.** alpha particles   **c.** neutrons   **d.** aluminum

7. Beta particles are ____ than alpha particles.
   a. larger   **b.** faster   **c.** slower   **d.** less powerful

8. A ____ can stop alpha particles.
   a. block of concrete   **c.** sheet of paper
   b. sheet of aluminum   **d.** photographic plate

9. An alpha particle is made of ____.
   a. two protons and two neutrons   **c.** gamma rays
   b. two electrons and two neutrons   **d.** two atoms

10. When an atom decays, it gives off or ____ particles or energy.
   a. stops   **b.** finds   **c.** emits   **d.** breaks down

Ⓒ COMPLETION Complete the sentences. *(20 points: 4 points each)*

Example: The time it takes for half the atoms in a sample to decay is its ____half-life____.

11. Only an _____ source can make an image on a photographic plate.

12. Doctors use radiation to clean _____ used in surgery.

13. Materials that are _____ give off energy and particles.

14. Alpha _____ are slow and large.

15. A smoke detector contains a small amount of _____ material.

Name _____ Date _____

PHYSICAL SCIENCE Radiation and Radioactivity

D **SHORT ANSWERS** Look at the table. Answer the questions.
(20 points: 4 points each)

Element	Half-Life	Use
carbon-14	5,730 years	We use it to find out how old a fossil or an artifact is.
cobalt-60	5.3 years	We use it to sterilize surgical instruments.
uranium-238	4.47 billion years	We use it as fuel for nuclear power.

Example: Which element has the shortest half-life?

_____ *cobalt-60* _____

16. Which element can be used to find the age of a fossil?

   _____

17. Which element has a half-life that is billions of years?

   _____

18. Where is uranium-238 used?

   _____

19. How can we use cobalt-60?

   _____

20. What is the half-life of carbon-14?

   _____

E **WRITING** Why do we remember Antoine-Henri Becquerel today? How did he make his discovery? Why was his discovery important? Write a paragraph.
(20 points)

_____

_____

_____

_____

_____

GATEWAY TO SCIENCE Assessment Book • Copyright © Thomson Heinle

110

Name _____ Date _____

Student book pages 186–189

A TRUE/FALSE Write if the sentence is true (T) or false (F). If the sentence is false, change the underlined word or phrase to make it true. *(20 points: 4 points each)*

Example: Friction always works against gravity. __F__ ____motion____

1. Objects with more **mass** have more gravity. ____ _____

2. **Balanced** forces cause a change in motion. ____ _____

3. *Push* and *pull* mean **the opposite of** each other. ____ _____

4. **Friction** can cause heat. ____ _____

5. Gravity is a **motion**. ____ _____

B MULTIPLE CHOICE Choose the correct answer. *(20 points: 4 points each)*

Example: A pull changes an object's __c__.
    a. mass   **b.** gravity   **c.** motion   **d.** weight

6. The sun's ____ pulls on the planets.
    a. mass   **b.** weight   **c.** gravity   **d.** friction

7. The measure of the pull of gravity on an object is its ____.
    a. mass   **b.** equilibrium   **c.** friction   **d.** weight

8. A newton is a measure of ____.
    a. weight   **b.** mass   **c.** friction   **d.** motion

9. Earth has more ____ than the moon.
    a. friction   **b.** equilibrium   **c.** mass   **d.** balanced forces

10. Friction makes it ____ to push a box across a rough surface.
    a. faster   **b.** unbalanced   **c.** harder   **d.** easier

C COMPLETION Complete the sentences. *(20 points: 4 points each)*

Example: Friction happens when surfaces ____run against____ each other.

11. Earth has more gravity than the moon because it has more _____.

12. Gravity is the force of _____ between objects.

13. Earth's gravity makes a skateboarder go _____ down a hill.

14. The sun's _____ keeps the planets in their orbits.

15. Forces that are in _____ are balanced.

D **SHORT ANSWERS** Look at the table. Answer the questions.
(20 points: 4 points each)

Planet	Weight of an Average Person in Newtons (N)
Mercury	302.4 N
Venus	725.6 N
Earth	800 N
Mars	301.6 N
Jupiter	2,026.4 N
Saturn	851.2 N
Uranus	711.2 N
Neptune	900 N

Example: On which planet does a person weigh the most?

       Jupiter

16. On which planet does a person weigh the least?

_____

17. Which planet's gravity is most like Earth's gravity?

_____

18. Which planet's gravity is most like Venus's gravity?

_____

19. You fly from Mars to Neptune. Do you weigh more or less?

_____

20. Which planet has the least gravity? _____

E **WRITING** Your class is having a tug-of-war. The flag is not moving. Then your teacher joins one side. Explain what happens and why. Write a paragraph.
(20 points)

_____

_____

_____

_____

Name _____ Date _____

Grade

A **TRUE/FALSE** Write if the sentence is true (T) or false (F). If the sentence is false, change the underlined word or phrase to make it true. *(20 points: 4 points each)*

Example: Action and reaction forces occur in **the same direction**. __F__

opposite directions

1. The acceleration of an object is related to its **mass**. ____ _____

2. How fast something is moving is its **velocity**. ____ _____

3. More **friction** creates more acceleration. ____ _____

4. Forces always happen **in pairs**. ____ _____

5. Isaac Newton explained **four** laws of motion. ____ _____

B **MULTIPLE CHOICE** Choose the correct answer. *(20 points: 4 points each)*

Example: Action and reaction forces are always __b__.
   a. at rest   **b.** equal   **c.** accelerating   **d.** resisting change

6. Because of ____ an object stays at rest until a force affects it.
   a. velocity   **b.** acceleration   **c.** speed   **d.** inertia

7. Acceleration is the rate at which ____ changes.
   a. friction   **b.** inertia   **c.** velocity   **d.** force

8. Speed in a direction is ____.
   a. acceleration   **b.** velocity   **c.** friction   **d.** inertia

9. Acceleration ____ as an object comes to a stop.
   a. increases   **b.** is equal   **c.** reacts   **d.** decreases

10. Meters per second is a measure of ____.
   a. speed   **b.** inertia   **c.** velocity   **d.** friction

C **COMPLETION** Complete the sentences. *(20 points: 4 points each)*

Example: An object in motion will stay in motion until _a force stops it_.

11. Meters per second per second is a measure of _____.

12. When a car suddenly stops, the passengers keep _____.

13. When an object's speed or direction changes, its _____ changes.

14. One object has more mass than another object. You need to use more _____ to move the object with more mass.

15. You roll a ball. Its _____ depends on how much force you use.

D **SHORT ANSWERS** Look at the formula. Answer the questions.
(20 points: 4 points each)

$$F \text{ (force)} = m \text{ (mass)} \times a \text{ (acceleration)}$$

Example: What two things affect force? ___*mass and acceleration*___

16. What is the force in newtons if the mass is 2 kg and the acceleration is 10 m/s/s?

   _____

17. What is the acceleration in meters per second per second if the force is 55 newtons and the mass is 11 kg? _____

18. What is the mass in kilograms if the force is 63 newtons and the acceleration is 9 m/s/s? _____

19. What happens if you double the amount of force affecting a certain mass?

   _____

20. What happens if you apply half as much force to a certain amount of mass?

   _____

E **WRITING** A batter hits a baseball, and a fielder catches it. Describe the forces that affect the ball. Write a paragraph. *(20 points)*

_____

_____

_____

_____

_____

GATEWAY TO SCIENCE Assessment Book • Copyright © Thomson Heinle

Name _____ Date _____

Student book pages 194–197

Grade

A **TRUE/FALSE** Write if the sentence is true (T) or false (F). If the sentence is false, change the <u>underlined</u> word or phrase to make it true. *(20 points: 4 points each)*

Example: A lever changes the **distance** of a force. ___F___ ____directions____

1. A compound machine makes work **greater** than a simple machine makes work.

   ____ _____

2. An inclined plane makes it **easier** to raise a heavy object. ____ _____

3. A screw is a **simple machine**. ____ _____

4. You can use a **wedge** to split wood. ____ _____

5. A shovel is a **compound machine**. ____ _____

B **MULTIPLE CHOICE** Choose the correct answer. *(20 points: 4 points each)*

Example: A wedge changes the __b__ of force.
   a. distance   **b.** direction   **c.** work   **d.** load

6. A lever is a board or rod that turns on a ____.
   a. pulley   **b.** wedge   **c.** load   **d.** fulcrum

7. Force multiplied by distance equals ____.
   a. work   **b.** effort   **c.** load   **d.** direction

8. The handle of a shovel is ____.
   a. a wedge   **c.** a lever
   b. an inclined plane   **d.** a screw

9. A machine can help you lift ____.
   a. energy   **b.** a fulcrum   **c.** a load   **d.** a force

10. The part of a shovel that scoops snow is a ____.
   a. fulcrum   **b.** lever   **c.** pulley   **d.** wedge

C **COMPLETION** Complete the sentences. *(20 points: 4 points each)*

Example: An inclined plane spreads work over a greater ____distance____.

11. A seesaw uses a _____ force to lift a person.

12. A wedge changes the _____ of a force.

13. A compound machine is made of more than one _____.

14. A wheel and axle is a type of _____.

15. Scissors are a kind of _____ machine.

Name _____ Date _____

D **SHORT ANSWERS** Look at the illustration. Answer the questions.
(20 points: 4 points each)

Example: Which person is using less force?

_____person A_____

16. Which person is moving the box a shorter distance?

_____

17. Is one person doing more work? _____

18. Person A makes the inclined plane longer. Does this increase or decrease the amount
of force needed to move the box? _____

19. Person A makes the inclined plane shorter. How does this affect the amount of work
done? _____

20. Person B lifts the box higher than person A. Does person B do more or less work
than person A? _____

E **WRITING** You have to move some rocks in the garden. You can carry them in a
basket or wheel them in a wheelbarrow. Which will take less force? Why? Will the
amount of work change? Write a paragraph. *(20 points)*

_____

_____

_____

_____

_____

GATEWAY TO SCIENCE Assessment Book • Copyright © Thomson Heinle

Name _____ Date _____

Grade ☐

A **TRUE/FALSE** Write if the sentence is true (T) or false (F). If the sentence is false, change the <u>underlined</u> word or phrase to make it true. *(20 points: 4 points each)*

Example: The trough is the **highest** part of a water wave. __F__ _____lowest_____

1. Long wavelengths have **high** energy. ____ _____

2. Sound waves are **transverse** waves. ____ _____

3. **Wavelength** is the distance between the crests of a wave. ____ _____

4. **Sound waves** can move through steel. ____ _____

5. We **can see** part of the electromagnetic spectrum. ____ _____

B **MULTIPLE CHOICE** Choose the correct answer. *(20 points: 4 points each)*

Example: The particles in a water wave move __b__.
   a. back and forth   **b.** up and down   **c.** forward   **d.** through the air

6. Radio waves are ____.
   a. longitudinal waves   **c.** electromagnetic waves
   b. sound waves   **d.** transverse waves

7. The highest part of a wave is the ____.
   a. crest   **b.** resting point   **c.** wavelength   **d.** amplitude

8. What kind of wave can travel through empty space? ____
   a. longitudinal wave   **c.** sound wave
   b. transverse wave   **d.** electromagnetic wave

9. Particles in a ____ move back and forth.
   a. longitudinal wave   **c.** water wave
   b. electromagnetic wave   **d.** transverse wave

10. We can see ____.
   a. radio waves   **c.** visible light waves
   b. microwaves   **d.** X-rays

C **COMPLETION** Complete the sentences. *(20 points: 4 points each)*

Example: Water waves are _____transverse_____ waves.

11. The distance from trough to trough of a wave is the _____.

12. Doctors use _____ to see inside people.

13. In _____ waves, some particles are pushed together and some are spread apart.

14. The _____ of transverse waves is the distance from the resting place to the crest.

15. We use _____ waves to carry radio and TV signals.

D **SHORT ANSWERS** Look at the models. Answer the questions.
(20 points: 4 points each)

1.

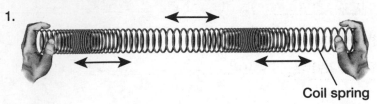

Coil spring

2.

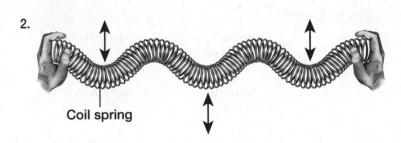

Coil spring

Example: In which directions do the particles move in the longitudinal wave?
   back and forth

16. What kind of wave is number (1)? _____

17. What kind of wave is number (2)? _____

18. What kind of wave is modeled by an up-and-down motion?

   _____

19. Which of the waves modeled has a crest and a trough?

   _____

20. Which of the waves modeled has compression and rarefaction?

   _____

E **WRITING** Name three kinds of waves that you used yesterday. What did you use them to do? *(20 points)*

_____

_____

_____

_____

_____

Name _____ Date _____

📖 Student book pages 202–205

Grade ☐

A **TRUE/FALSE** Write if the sentence is true (T) or false (F). If the sentence is false, change the underlined word or phrase to make it true. *(20 points: 4 points each)*

Example: People make **natural** light. ___F___ ___artificial___

1. All light passes through **opaque** objects. ____ _____

2. Smooth, shiny surfaces **refract** the most light. ____ _____

3. Light is **energy**. ____ _____

4. We see the world because of **light**. ____ _____

5. *Reflect* and *absorb* mean **the same thing**. ____ _____

B **MULTIPLE CHOICE** Choose the correct answer. *(20 points: 4 points each)*

Example: A ___b___ is an artificial light source.
  a. star   **b.** candle   **c.** prism   **d.** moon

6. Light bounces off a ____.
  a. prism   **b.** lamp   **c.** pencil   **d.** mirror

7. The window of a car is ___.
  a. translucent   **b.** opaque   **c.** transparent   **d.** reflected

8. A black object ____ all colors of light.
  a. absorbs   **b.** reflects   **c.** refracts   **d.** bends

9. Smooth, shiny surfaces ____ the most light.
  a. refract   **b.** reflect   **c.** absorb   **d.** separate

10. Frosted glass is ___.
  a. transparent   **b.** opaque   **c.** translucent   **d.** reflected

C **COMPLETION** Complete the sentences. *(20 points: 4 points each)*

Example: A white object ___reflects___ all colors of light.

11. A white object reflects all _____ of light.

12 A _____ divides light into its colors.

13. A pencil is in a glass of water. It looks bent because of _____.

14. In a prism, each _____ of light bends differently.

15. No light passes through a material. The material is _____.

D **SHORT ANSWERS** Look at the diagrams. Answer the questions.
(20 points: 4 points each)

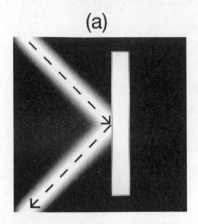
(a)

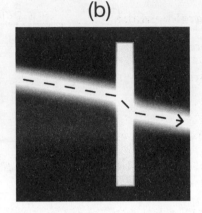
(b)

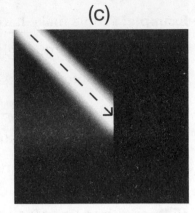
(c)

Example: What happens when light strikes the object in (a) above?

_____It is reflected._____

16. What words can describe the surface of the object in (a) above?

_____

17. What happens when light strikes the object in (b) above?

_____

18. What kind of object is in (b) above?

_____

19. What happens when light strikes the object in (c) above?

_____

20. What kind of object is in (c) above?

_____

E **WRITING** You got a new bike. Describe the bike. Explain why it looks like a certain color or colors. Tell whether any parts are transparent or translucent. Explain which parts reflect the most light. Write a paragraph. *(20 points)*

_____

_____

_____

_____

_____

GATEWAY TO SCIENCE Assessment Book • Copyright © Thomson Heinle

Name _____ Date _____

Grade

A TRUE/FALSE Write if the sentence is true (T) or false (F). If the sentence is false, change the <u>underlined</u> word or phrase to make it true. *(20 points: 4 points each)*

Example: Wind energy is **nonrenewable**. __F__ _____renewable_____

1. A dam is a source of **hydropower**. ____ _____

2. **Thermal** energy powers lights in your home. ____ _____

3. **Thermal energy** makes you warm. ____ _____

4. **Thermal energy** lets you see. ____ _____

5. Coal is a **renewable** energy source. ____ _____

B MULTIPLE CHOICE Choose the correct answer. *(20 points: 4 points each)*

Example: The ability to do work is __c__.
   a. thermal energy   **c.** mechanical energy
   b. chemical energy   **d.** hydropower

6. Sound energy lets you ____.
   a. see   **b.** hear   **c.** do work   **d.** stay warm

7. Kinetic energy is a form of ____.
   a. light energy   **c.** potential energy
   b. sound energy   **d.** mechanical energy

8. A soccer ball at the top of a steep hill has ____.
   a. kinetic energy   **b.** potential energy   **c.** sound energy   **d.** wind energy

9. Fossil fuels are ____ energy sources.
   a. renewable   **b.** kinetic   **c.** nonrenewable   **d.** mechanical

10. A stretched rubber band has stored ____.
   a. kinetic energy   **b.** potential energy   **c.** hydropower   **d.** sound energy

C COMPLETION Complete the sentences. *(20 points: 4 points each)*

Example: When you plug in a lamp, you are using ____electrical____ energy.

11. The battery in your wristwatch uses _____ energy.

12. A fire gives off light energy and _____ energy.

13. As you ride your bike down a hill you have _____ energy.

14. Bones in your ear vibrate because of _____ energy.

15. You can easily replace _____ energy sources.

D **SHORT ANSWERS**   Read the information. Answer the questions.
(20 points: 4 points each)

Visualize a person on a skateboard. The skateboard is at the top of a steep hill. Then visualize the person riding the skateboard down the hill.

Example: When did the person have potential energy?

_____at the top of the hill_____

16. When did the skateboard have potential energy?

_____

17. When did the person have kinetic energy?

_____

18. When did the skateboard have kinetic energy?

_____

19. Did the person have mechanical or electrical energy at the top of the hill?

_____

20. Did the skateboard have mechanical energy or light energy as it went down the hill?

_____

E **WRITING**   Your family had a picnic in the back yard. You lit candles. You cooked hamburgers over a fire in the grill. You drank cold drinks from the refrigerator. You listened to music on your portable CD player. What forms of energy did you use? Write a paragraph. *(20 points)*

_____

_____

_____

_____

_____

GATEWAY TO SCIENCE Assessment Book · Copyright © Thomson Heinle

Name _____ Date _____

Grade

A TRUE/FALSE Write if the sentence is true (T) or false (F). If the sentence is false, change the underlined word or phrase to make it true. *(20 points: 4 points each)*

Example: The materials in a battery have **thermal energy**. __F__ _chemical energy_

1. A thermometer measures **heat energy**. ____ _____

2. **Kinetic energy** shoots fireworks into the sky. ____ _____

3. Heat travels from hot chocolate to a spoon by **radiation**. ____ _____

4. The thermal energy from a toaster feels **cold**. ____ _____

5. Heat always moves **from warmer matter to cooler matter**. ____ _____

B MULTIPLE CHOICE Choose the correct answer. *(20 points: 4 points each)*

Example: A toaster uses __b__.
   a. kinetic energy   b. electrical energy   c. chemical energy   d. radiation

6. A flashlight changes chemical energy into ____.
   a. kinetic energy   b. radiation   c. light energy   d. sound energy

7. When a cold egg touches a hot pan, ____ moves to the egg.
   a. light energy   b. sound energy   c. kinetic energy   d. thermal energy

8. Radiation is heat transferred in ____.
   a. waves   b. particles   c. liquids   d. solids

9. A swinging pendulum always has ____ energy.
   a. potential   c. the same amount of
   b. kinetic   d. different amounts of

10. A toaster converts or ____ electrical energy into heat energy.
   a. moves   b. transforms   c. destroys   d. travels

C COMPLETION Complete the sentences. *(20 points: 4 points each)*

Example: The law of _conservation of energy_ says that energy cannot be made or destroyed.

11. A bicycle at the top of a steep hill has _____.

12. Conduction is the transfer of _____ from fast-moving particles to slower-moving particles.

13. A battery-powered radio changes chemical energy into _____.

14. The total _____ in a pendulum always stays the same.

15. The word part *trans* in the word *transfer* means _____.

Name _____ Date _____

D **SHORT ANSWERS** Look at the drawing. Answer the questions.
(20 points: 4 points each)

Pan

Convection

Hot plate

OFF   HOT

Example: What kind of movement does convection cause in the water?

_____ a circular movement _____

16. What kind of energy does the hot plate use?

_____

17. What kind of energy does the hot plate produce?

_____

18. How does thermal energy move from the hot plate to the pan?

_____

19. Which part of the water gets hot first?

_____

20. How does thermal energy move through the water?

_____

E **WRITING** You have toast and fried eggs for breakfast. What kind of energy do you use? How does the energy move from the toaster to the bread and from the pan to the eggs? Write a paragraph. *(20 points)*

_____

_____

_____

_____

_____

GATEWAY TO SCIENCE Assessment Book • Copyright © Thomson Heinle

A **TRUE/FALSE** Write if the sentence is true (T) or false (F). If the sentence is false, change the underlined word or phrase to make it true. *(20 points: 4 points each)*

Example: Chlorophyll traps **chemical energy**. __F__ ___light energy___

1. Plants and animals use energy to build new **cells**. _____ _____

2. Glucose is a kind of **plant cell**. _____ _____

3. Glucose contains **chemical energy**. _____ _____

4. All living things need **food**. _____ _____

5. Plants take energy **from the sun**. _____ _____

B **MULTIPLE CHOICE** Choose the correct answer. *(20 points: 4 points each)*

Example: Chlorophyll is found in __b__.
   a. glucose   **b.** chloroplasts   **c.** the atmosphere   **d.** the sun

6. Plants make food _____.
   a. during photosynthesis   **c.** in mitochondria
   b. during cellular respiration   **d.** from chemical energy

7. Without the sun, Earth would be too _____ for living things.
   a. light   **b.** hot   **c.** cold   **d.** dry

8. The sun provides or _____ energy that living things need.
   a. traps   **b.** stores   **c.** changes   **d.** supplies

9. Plants use light energy, water, and _____ to make food.
   a. oxygen   **b.** glucose   **c.** carbon dioxide   **d.** chemical energy

10. Sugar contains _____.
   a. light energy   **b.** chemical energy   **c.** water   **d.** mitochondria

C **COMPLETION** Complete the sentences. *(20 points: 4 points each)*

Example: In cellular respiration, ___chemical energy___ changes into the energy to do work.

11. Cells use chemical energy during _____ respiration.

12. Plants store _____ energy.

13. During cellular respiration, cells use chemical energy and _____ to make other kinds of energy.

14. Most living things depend on plants for _____.

15. The green matter found in plants is _____.

Name _____ Date _____

D SHORT ANSWERS Look at the diagram. Answer the questions.
(20 points: 4 points each)

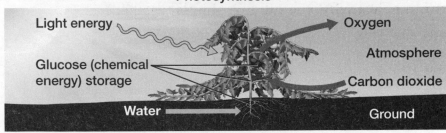

Photosynthesis

Light energy
Oxygen
Atmosphere
Glucose (chemical energy) storage
Carbon dioxide
Water
Ground

Example: What process does this diagram show?

_____ photosynthesis _____

16. Where does the light energy in the diagram come from?

_____

17. Where does the carbon dioxide in the diagram come from?

_____

18. Where does the water in the diagram come from?

_____

19. Where does the oxygen in the diagram come from?

_____

20. What kind of energy does the plant store?

_____

E WRITING You eat an apple. You get some energy. Explain how the energy got into the apple. Write a paragraph. *(20 points)*

_____

_____

_____

_____

_____

GATEWAY TO SCIENCE Assessment Book · Copyright © Thomson Heinle

Grade

A   TRUE/FALSE Write if the sentence is true (T) or false (F). If the sentence is false, change the <u>underlined</u> word or phrase to make it true. *(20 points: 4 points each)*

Example: To make **current electricity**, you rub two objects together. __F__

_____static electricity_____

1. Copper is a **conductor**. ____ _____

2. A flashlight contains a **battery**. ____ _____

3. **Current electricity** flows along a path. ____ _____

4. **Static electricity** jumps from place to place. ____ _____

5. Lightning is **static electricity**. ____ _____

B   MULTIPLE CHOICE Choose the correct answer. *(20 points: 4 points each)*

Example: The area that a magnet affects is its __b__.
    a. conductor   **b.** magnetic field   **c.** electromagnet   **d.** current

6. An electric charge is caused by ____ moving between atoms.
    a. copper   **b.** a magnetic field   **c.** a circuit   **d.** electrons

7. Plastic is ____.
    a. an insulator   **b.** a conductor   **c.** a circuit   **d.** magnetic

8. An electric current is a charge that flows along ____.
    a. plastic   **b.** a magnet   **c.** a circuit   **d.** a switch

9. Metal wires spinning inside ____ make an electric current.
    a. a magnetic field   **c.** an electric generator
    b. an electromagnet   **d.** an atom

10. An electromagnet attracts ____.
    a. metals   **b.** plastic   **c.** atoms   **d.** lightning

C   COMPLETION Complete the sentences. *(20 points: 4 points each)*

Example: A conductor is a material that can carry _____electricity_____.

11. The north pole of a magnet will attract _____ of another magnet.

12. An electromagnet uses electricity to create a _____.

13. In an _____, wires spin inside a magnetic field.

14. If you rub a balloon on a cat's fur, you create _____ electricity.

15. An object has a negative electric charge when it gains _____.

Electricity and Magnetism

PHYSICAL SCIENCE

D **SHORT ANSWERS** Look at the diagram. Answer the questions.
(20 points: 4 points each)

A Flashlight Circuit Diagram

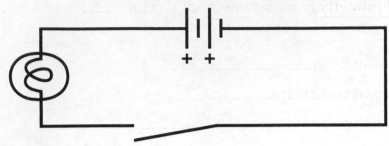

Example: What does this diagram show?

the circuit in a flashlight

16. What does the symbol that looks like a spring stand for?

_____

17. What do the lines with plus signs (++) stand for?

_____

18. What does the symbol that looks like a door stand for?

_____

19. What does the line that runs between the symbols represent?

_____

20. How do you know that electricity cannot flow along the circuit in the diagram?

_____

E **WRITING** You rub two objects together. What kind of electricity do you make?
Explain how this happens. Write a paragraph. *(20 points)*

_____

_____

_____

_____

_____

Name _____ Date _____

Grade

A **TRUE/FALSE** Write if the sentence is true (T) or false (F). If the sentence is false, change the <u>underlined</u> word or phrase to make it true. *(20 points: 2 points each)*

Example: Taste is a **chemical property** of matter. __F__ __physical property__

1. Matter has **two** states. ____ _____

2. Neutrons are **smaller** than atoms. ____ _____

3. An **endothermic** reaction gives off energy. ____ _____

4. Objects having more **mass** have more gravity. ____ _____

5. A compound machine makes work **greater** than a simple machine makes work.

   ____ _____

6. All light passes through **translucent** objects. ____ _____

7. A thermometer measures **chemical energy**. ____ _____

8. Copper is a **conductor**. ____ _____

9. A liquid has a definite **shape**. ____ _____

10. Different materials have different masses for the same **weight**.

    ____ _____

B **MULTIPLE CHOICE** Choose the correct answer. *(40 points: 2 points each)*

Example: You can see the __a__ of an object.
   a. color   **b.** odor   **c.** smell   **d.** taste

11. The particles in ____ are very far away from one another.
    a. a solid   **b.** a liquid   **c.** a gas   **d.** water

12. When you put a solid in a liquid, ____ pushes up on the object.
    a. density   **b.** force of gravity   **c.** buoyant force   **d.** mass

13. The particles that circle an atom's nucleus are ____.
    a. protons   **b.** electrons   **c.** neutrons   **d.** molecules

14. When wood burns, a ____ happens.
    a. physical change   **c.** chemical property
    b. pure substance   **d.** chemical change

B **MULTIPLE CHOICE,** continued

15. Every compound has ____.

    a. a symbol   **b.** a formula   **c.** a solution   **d.** an equation

16. Doctors use ____ to check for broken bones.

    a. X-rays   **b.** alpha particles   **c.** neutrons   **d.** molecules

17. The sun's ____ pulls on the planets.

    a. mass   **b.** weight   **c.** gravity   **d.** orbit

18. Because of ____ an object stays at rest until a force affects it.

    a. mass   **b.** acceleration   **c.** speed   **d.** inertia

19. A lever is a board or rod that turns on a ____.

    a. wheel   **b.** wedge   **c.** load   **d.** fulcrum

20. Radio waves are ____.

    a. longitudinal waves   **c.** electromagnetic waves
    b. sound waves   **d.** water waves

21. Light bounces off a ____.

    a. prism   **b.** lamp   **c.** chair   **d.** mirror

22. Sound energy lets you ____.

    a. see   **b.** hear   **c.** do work   **d.** taste

23. A flashlight changes chemical energy into ____.

    a. kinetic energy   **b.** potential energy   **c.** light energy   **d.** sound energy

24. Plants make food ____.

    a. during photosynthesis   **c.** in mitochondria
    b. during cellular respiration   **d.** in their roots

25. An electric charge is caused by ____ moving between atoms.

    a. copper   **b.** electromagnets   **c.** a circuit   **d.** electrons

26. Each ____ has a box on the periodic table.

    a. molecule   **b.** proton   **c.** element   **d.** neutron

27. A pure substance is made of one kind of ____.

    a. solution   **b.** mixture   **c.** formula   **d.** particle

GATEWAY TO SCIENCE Assessment Book • Copyright © Thomson Heinle

Name _____ Date _____

B MULTIPLE CHOICE, continued

28. Iron and oxygen combine to form ____.
    a. sodium dioxide   **b.** water   **c.** chlorine   **d.** rust

29. Beta particles are ____ than alpha particles.
    a. larger   **b.** faster   **c.** slower   **d.** heavier

30. The measure of the pull of gravity on an object is its ____.
    a. mass   **b.** balance   **c.** friction   **d.** weight

C COMPLETION Complete the sentences. *(20 points: 2 points each)*

Example: We measure boiling points with a ____thermometer____.

31. The _____ of water is 100°C.

32. During a _____ change, the identity of the substance does not change.

33. Only an _____ source can make an image on a photographic plate.

34. Meters per second per second is a measure of _____.

35. The distance from crest to crest of a wave is the _____.

36. The battery in a cell phone uses _____ energy.

37. Cells use chemical energy during _____ respiration.

38. Doctors use _____ to see bones inside people.

39. A fire gives off light energy and _____ energy.

40. Plants store _____ energy.

D **SHORT ANSWERS** Look at the formula. Answer the questions.
(20 points: 2 points each)

$$F \text{ (force)} = m \text{ (mass)} \times a \text{ (acceleration)}$$

Example: What two things affect force? _____ *mass and acceleration* _____

41. What is the force in newtons if the mass is 2 kg and the acceleration is 10 m/s/s?

_____

42. What is the acceleration in meters per second per second if the force is 55 newtons
and the mass is 11 kg? _____

43. What is the mass in kilograms if the force is 63 newtons and the acceleration is

9 m/s/s? _____

44. What happens when you double the amount of force affecting a certain mass?

_____

45. What happens when you apply half as much force to a certain amount of mass?

_____

Name _____   Date _____

D SHORT ANSWERS, continued

Look at the diagram. Answer the questions.

46. What kind of energy does the hot plate use?

    _____

47. What kind of energy does the hot plate produce?

    _____

48. How does thermal energy move from the hot plate to the pan?

    _____

49. Which part of the water gets hot first?

    _____

50. How does thermal energy move through the water?

    _____

E WRITING ASSESSMENT Write paragraphs. *(100 points: 25 points each)*

51. Water has three states. Name each state. Describe its physical properties. Tell about each state's shape, volume, and particles.

_____

_____

_____

_____

_____

52. A seesaw is one type of simple machine. Describe the parts of a seesaw. Explain how it makes work easier.

_____

_____

_____

_____

_____

53. Sunlight shines through raindrops. You see a rainbow. Explain how this occurs.

_____

_____

_____

_____

_____

54. How can you use electricity to make a magnet? How can you use a magnet to make electricity?

_____

_____

_____

_____

_____

GATEWAY TO SCIENCE Assessment Book • Copyright © Thomson Heinle

📖 Student book pages 002–221

Grade

A **TRUE/FALSE** Write if the sentence is true (T) or false (F). If the sentence is false, change the underlined word or phrase to make it true. *(20 points: 2 points each)*

Example: Robert Hook looked at cells through a **telescope**. __F__ _____microscope_____

1. A Venn diagram contains **one square**. ____ _____

2. A multicellular animal contains **two kinds** of cell. ____ _____

3. A fruit may contain one or more **spores**. ____ _____

4. Photosynthesis produces **oxygen** as waste. ____ _____

5. The Milky Way contains about 100 billion **galaxies**. ____ _____

6. A supergiant is much **larger** than the sun. ____ _____

7. A **meteoroid** can be as small as a piece of dust. ____ _____

8. A **property** is a characteristic or trait. ____ _____

9. A compound is a **pure substance**. ____ _____

10. Atoms can join together to form **molecules**. ____ _____

B **MULTIPLE CHOICE** Choose the correct answer. *(40 points: 2 points each)*

Example: The cell membrane of a bacteria cell is __d__ .
    a. jelly-like   **b.** stiff   **c.** empty   **d.** thick

11. A microscope helps us see ____.
    a. small objects   **c.** things that are in a graduated cylinder
    b. things that are far away   **d.** things in a computer

12. When you prevent accidents, you ____.
    a. help them   **b.** harm them   **c.** avoid them   **d.** protect them

13. Most animals have ____.
    a. senses   **b.** outer shells   **c.** suckers   **d.** skeletons

14. A beetle has ____.
    a. a hard body case   **c.** a cocoon
    b. a backbone   **d.** a chrysalis

B MULTIPLE CHOICE, continued

15. A crocodile is a kind of ____.

   a. amphibian   **b.** mammal   **c.** snake   **d.** reptile

16. The ____ is the main part of the circulatory system.

   a. trachea   **b.** lung   **c.** kidney   **d.** heart

17. A strawberry plant uses ____ to reproduce.

   a. bulbs   **b.** runners   **c.** yeast   **d.** fruit

18. During crossing over, small sections of chromosomes ____.

   a. form a nucleus   **c.** switch places
   b. divide into daughter cells   **d.** move apart

19. On Earth, one ____ is 24 hours long.

   a. night and day   **b.** orbit   **c.** year   **d.** month

20. The sun, moon, and Earth line up during ____.

   a. a full moon   **b.** a week   **c.** a tidal range   **d.** a neap tide

21. Rock breaks into tiny pieces because of ____.

   a. weathering   **b.** volcanoes   **c.** lava   **d.** heat

22. Oxygen and carbon dioxide can cause ____.

   a. mechanical weathering   **c.** glaciers
   b. chemical weathering   **d.** moraines

23. When a tornado forms over water, it is called a ____.

   a. hurricane   **b.** cyclone   **c.** wall cloud   **d.** waterspout

24. Something that is *beneath* the ocean is ____ the ocean.

   a. above   **b.** apart   **c.** under   **d.** against

25. A ____ can stop alpha particles.

   a. block of concrete   **c.** sheet of paper
   b. sheet of aluminum   **d.** an umbrella

26. A newton is a measure of ____.

   a. weight   **b.** mass   **c.** friction   **d.** tons

Name _____   Date _____

B MULTIPLE CHOICE, continued

27. The handle of a shovel is a ____.

    a. wedge   **b.** inclined plane   **c.** lever   **d.** fulcrum

28. What kind of wave can travel through empty space? ____

    a. longitudinal wave   **c.** water wave

    b. transverse wave   **d.** electromagnetic wave

29. A black object ____ all colors of light.

    a. absorbs   **b.** reflects   **c.** refracts   **d.** transmits

30. A soccer ball at the top of a steep hill has ____.

    a. kinetic energy   **c.** sound energy

    b. potential energy   **d.** water energy

C COMPLETION Complete the sentences. *(20 points: 2 points each)*

Example: A bacteria cell has a cell membrane and ____cytoplasm____.

31. A scientist studies a bar graph and finds out which material keeps ice frozen longest.

    The scientist is making a _____.

32. In binomial nomenclature, the name's second part is the _____.

33. In parasitism, one organism in the relationship is harmed and the other is

    _____.

34. Some _____ open only in the morning.

35. Astronauts make repairs and do experiments while living on _____.

36. Animals get _____ from water or from the air.

37. Snow and rain are types of _____.

38. During cellular respiration, cells use _____ and chemical energy to

    make other kinds of energy.

39. An electromagnet uses electricity to create a _____.

40. In an _____, wires spin inside a magnetic field.

Name _____ Date _____

D SHORT ANSWERS Look at the thermometers. Answer the questions.
(20 points: 2 points each)

Example: At what temperature does water freeze on the Celsius scale?

_____ zero degrees _____

41. At what temperature does water boil on the Fahrenheit scale?

_____

42. At what temperature does water boil on the Celsius scale?

_____

43. At what temperature does water freeze on the Fahrenheit scale?

_____

44. Which temperature scale has more degrees between the freezing point and the

boiling point of water? _____

45. The temperature today is 68 degrees Fahrenheit. What is the temperature in degrees

Celsius? _____

Name _____ Date _____

D **SHORT ANSWERS,** continued

Look at the pie chart. Answer the questions.

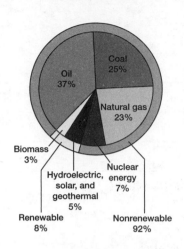

46. How much of the energy comes from renewable resources?

_____

47. How much more energy comes from oil than from biomass?

_____

48. What percentage of energy comes from nonrenewable resources?

_____

49. How much less energy comes from renewable resources than from coal?

_____

50. What percent of energy is provided by fossil fuels (coal, oil, and natural gas)?

_____

E **WRITING ASSESSMENT** Write paragraphs. *(100 points: 25 points each)*

51. You are a science teacher. What safety equipment must your students wear in the lab? Why do they need each piece of equipment?

_____

_____

_____

_____

_____

52. How are an animal cell and a plant cell different? How are they alike? Describe their parts.

_____

_____

_____

_____

_____

53. What is a supergiant? How big is it? What can it become? How can its life end?

_____

_____

_____

_____

_____

54. What kinds of atoms make up a water molecule? How many atoms are in water? How is water different from the elements it is formed from?

_____

_____

_____

_____

_____

GATEWAY TO SCIENCE Assessment Book • Copyright © Thomson Heinle

RUBRIC FOR WRITING ASSESSMENT

Use this rubric to score students' writing on quizzes and end-of-section tests.

The rubric is based on a perfect score of 20 points (quizzes) and 25 points (end-of-section tests).

Elements of Good Writing	Quiz Points	End-of-section Test Points	Description
Organization	4 points	5 points	• Ideas progress logically • Ideas are supported and explained
Sentence Fluency	3 points	4 points	• Writing stays focused • Writing seems complete • Writing is meaningful and coherent
Word Choice	3 points	4 points	• Writer uses exact words to clarify and enhance meaning • Writer uses language effectively
Conventions	5 points	6 points	• Punctuation and capitalization are appropriate • Spelling errors are few or none • Grammar and usage are consistently appropriate • Words, phrases, and sentence structures are used correctly and effectively
Presentation	5 points	6 points	• Penmanship is pleasing • Margins and spacing are appropriate • Devices (headings, bullets, numbers, etc.) clarify meaning

ANSWER SHEET FOR QUIZZES

Name _____ Date _____

Grade

Fill in the circles of the correct answers and write your answers on the lines.

1. ____ _____

2. ____ _____

3. ____ _____

4. ____ _____

5. ____ _____

6. ⓐ ⓑ ⓒ ⓓ

7. ⓐ ⓑ ⓒ ⓓ

8. ⓐ ⓑ ⓒ ⓓ

9. ⓐ ⓑ ⓒ ⓓ

10. ⓐ ⓑ ⓒ ⓓ

11. _____

12. _____

13. _____

14. _____

15. _____

16. _____

17. _____

18. _____

19. _____

20. _____

ANSWER SHEET FOR END-OF-SECTION TESTS
AND END-OF-BOOK TEST

Name _____ Date _____

Grade

Fill in the circles of the correct answers and write your answers on the lines.

1. ____ _____   26. ⓐ ⓑ ⓒ ⓓ

2. ____ _____   27. ⓐ ⓑ ⓒ ⓓ

3. ____ _____   28. ⓐ ⓑ ⓒ ⓓ

4. ____ _____   29. ⓐ ⓑ ⓒ ⓓ

5. ____ _____   30. ⓐ ⓑ ⓒ ⓓ

6. ____ _____   31. _____

7. ____ _____   32. _____

8. ____ _____   33. _____

9. ____ _____   34. _____

10. ____ _____   35. _____

11. ⓐ ⓑ ⓒ ⓓ   36. _____

12. ⓐ ⓑ ⓒ ⓓ   37. _____

13. ⓐ ⓑ ⓒ ⓓ   38. _____

14. ⓐ ⓑ ⓒ ⓓ   39. _____

15. ⓐ ⓑ ⓒ ⓓ   40. _____

16. ⓐ ⓑ ⓒ ⓓ   41. _____

17. ⓐ ⓑ ⓒ ⓓ   42. _____

18. ⓐ ⓑ ⓒ ⓓ   43. _____

19. ⓐ ⓑ ⓒ ⓓ   44. _____

20. ⓐ ⓑ ⓒ ⓓ   45. _____

21. ⓐ ⓑ ⓒ ⓓ   46. _____

22. ⓐ ⓑ ⓒ ⓓ   47. _____

23. ⓐ ⓑ ⓒ ⓓ   48. _____

24. ⓐ ⓑ ⓒ ⓓ   49. _____

25. ⓐ ⓑ ⓒ ⓓ   50. _____

Thinking Like a Scientist (pp 1–2)

A TRUE/FALSE

1. T   2. T   3. F, predictions   4. T   5. T

B MULTIPLE CHOICE

6. a   7. c   8. b   9. d   10. c

C COMPLETION

11. observation   12. hypothesis   13. conclusion
14. correlational   15. experimental

D SHORT ANSWERS

16. seeds   17. thin, pointed   18. caterpillars
and flying insects   19. hummingbird
20. nectar from flowers

E WRITING

correlational design; write down what kind of home

each bird lives in; gather data on bird sizes; make a

chart; look for relationships

Science Tools (pp 3–4)

A TRUE/FALSE

1. F, satellites   2. T   3. F, volume   4. F, are not   5. T

B MULTIPLE CHOICE

6. c   7. b   8. d   9. c   10. a

C COMPLETION

11. telescope   12. fine adjustment knob
13. satellite   14. slide   15. measuring

D SHORT ANSWERS

16. the one on the right   17. the rock   18. the
bottom of the meniscus   19. at the top
20. the volume of the water

E WRITING

measure air temperature with a thermometer;

measure wind speed with an anemometer; Students

may answer "measure rain in a graduated cylinder."

Metric Units of Measurement (pp 5–6)

A TRUE/FALSE

1. T   2. F, thousand   3. F, Mars   4. T   5. T

B MULTIPLE CHOICE

6. d   7. b   8. c   9. d   10. b

C COMPLETION

11. length or distance   12. volume
13. three thousandths   14. thermometer
15. one thousandth or 0.001

D SHORT ANSWERS

16. 212 degrees   17. 100 degrees   18. 32 degrees
19. Fahrenheit scale   20. 20 degrees

144

E WRITING

The nurse uses kilograms to measure mass;

centimeters or meters to measure height; degrees

Celsius to measure temperature.

Data Analysis (pp 7–8)

A TRUE/FALSE

1. T   2. T   3. F, pie chart   4. F, two circles   5. T

B MULTIPLE CHOICE

6. c   7. d   8. b   9. c   10. a

C COMPLETION

11. relationship   12. across   13. line
14. sections or slices   15. 100

D SHORT ANSWERS

16. July and August   17. January   18. March
19. November and December   20. more

E WRITING

measure the temperature every day during May;

organize data by day of the month; put results in a

line graph; draw a graph that shows temperature

over time; connect the points with a line

Safety in the Lab (pp 9–10)

A TRUE/FALSE

1. F, first aid kit   2. T   3. F, eyes   4. F, wash the
area   5. T

B MULTIPLE CHOICE

6. d   7. a   8. c   9. b   10. c

C COMPLETION

11. living   12. soap   13. pictures
14. directions   15. safety signs

D SHORT ANSWERS

16. a sharp object   17. Corrosive Chemical
18. you could get shocked or hurt   19. Sharp
Object   20. goggles, gloves, and a lab apron

E WRITING

tell the teacher; wash the burn in cold water; use a

first aid kit to treat the burn

Science Basics End-of-Section Test
(pp 11–16)

A TRUE/FALSE

1. T   2. T   3. F, satellites   4. T   5. T
6. F, thousand   7. T   8. T   9. F, first aid kit   10. T

B MULTIPLE CHOICE

11. a   12. c   13. b   14. d   15. c   16. b
17. d   18. c   19. d   20. b   21. c   22. d

GATEWAY TO SCIENCE Assessment Book • Copyright © Thomson Heinle

23. c 24. d 25. b 26. c 27. d 28. a
29. c 30. b

C COMPLETION

31. observation 32. hypothesis 33. telescope
34. fine adjustment knob 35. length or distance
36. volume 37. relationship 38. across
39. living 40. soap

D SHORT ANSWERS

41. seeds 42. thin, pointed 43. caterpillars
and flying insects 44. hummingbird 45. nectar
from flowers 46. the one on the right 47. the
rock 48. the bottom of the meniscus 49. at the
top 50. the volume of the water

E WRITING ASSESSMENT

51. "Write down observations; form a hypothesis;
make a prediction; gather data; look for relationships;
do an experiment to see if the oil kills plants and
trees; make a conclusion; tell others about the
results" are three steps the scientist might take in
an investigation of why the trees died.

52. Fill a graduated cylinder half full of water, record
the volume; put in one object; record the volume;
subtract the volume of water; find volume of object.
Repeat this process with a second object in a second
cylinder. Compare volumes of the two objects.

53. People in different places used different
measurement systems; they didn't understand each
other; they couldn't tell each other the size of
something; they couldn't tell each other how far away
something was; with the metric system, they
understood each other.

54. Venn diagram; two circles that overlap; write
things that are true of both cats and dogs in center;
write things that are true of cats on one side; write
things that are only true of dogs on the other side.

The Cell (pp 17–18)

A TRUE/FALSE

1. T 2. F, cells 3. F, cell membrane 4. T 5. T

B MULTIPLE CHOICE

6. a 7. a 8. c 9. b 10. b

C COMPLETION

11. organelles 12. chloroplasts 13. bacteria
14. microscope 15. cell wall

D SHORT ANSWERS

16. animal or bacteria 17. 2 18. 2 19. 4
20. plant

E WRITING

Robert Hooke studied cork. He looked at cork under
a microscope. He saw its empty cell walls. He drew a
picture of what he saw. He invented the word cell.

Single-Celled Organisms (pp 19–20)

A TRUE/FALSE

1. F, water 2. F, chloroplasts 3. T 4. T 5. T

B MULTIPLE CHOICE

6. a 7. d 8. b 9. c 10. d

C COMPLETION

11. fungus 12. green 13. tube 14. bacterium
15. salty

D SHORT ANSWERS

16. chlorophyll 17. the amoeba 18. when
there is no sunlight 19. euglena and green algae
20. the amoeba

E WRITING

places with no air; such as inside a cow's stomach;
places deep in the ocean where the water is very
salty; inside tube worms where the water is very hot

Multicellular Organisms (pp 21–22)

A TRUE/FALSE

1. T 2. F, many kinds 3. F, chromosomes
4. T 5. F, white blood cells

B MULTIPLE CHOICE

6. c 7. d 8. a 9. a 10. c

C COMPLETION

11. leaves 12. disease 13. chromosomes
14. Lung 15. Skin

D SHORT ANSWERS

16. bone cells 17. smooth muscle cells
18. bone cells 19. help move food
20. red blood cells

E WRITING

A tissue is a body part made from many cells. All
the cells are the same kind. All the cells in a tissue
do the same kind of work. Lung tissue helps animals
breathe. Tubes that carry things between roots and
stems in plants are tissues.

Plants (pp 23–24)

A TRUE/FALSE

1. T 2. F, seeds 3. F, seeds 4. F, roots 5. T

B MULTIPLE CHOICE

6. b 7. a 8. c 9. b 10. a

C COMPLETION

11. nutrients   12. clay   13. nitrogen   14. loam
15. minerals

D SHORT ANSWERS

16. humus   17. 2 or 3%   18. water and air
19. about 47%   20. 50%

E WRITING

Like other plants, the Venus flytrap has leaves; it
needs nitrogen; the flytrap gets nitrogen from
insects; other plants get nitrogen from soil.

Kinds of Plants (pp 25–26)

A TRUE/FALSE

1. F, seed   2. F, pollen   3. T   4. F, seeds   5. F, does

B MULTIPLE CHOICE

6. d   7. c   8. c   9. d   10. a

C COMPLETION

11. pistils   12. pollination   13. roots
14. germinates   15. spores

D SHORT ANSWERS

16. it travels down a tube   17. inside the
bottom of the pistil   18. after   19. after
20. a seed develops

E WRITING

The seed absorbs water; the seed's hard outer shell
breaks; a root grows out of one end of the seed;
a stem grows out of the other end of the seed;
a seedling grows.

Photosynthesis (pp 27–28)

A TRUE/FALSE

1. T   2. T   3. F, roots   4. F, chloroplasts   5. T

B MULTIPLE CHOICE

6. d   7. d   8. c   9. a   10. b

C COMPLETION

11. rain   12. food   13. sugar   14. leaves
15. roots

D SHORT ANSWERS

16. carbon dioxide   17. oxygen   18. sunlight,
carbon dioxide, and water   19. photosynthesis
20. glucose

E WRITING

Count the tree's annual rings to tell its age; find last
year's rings by counting one ring from the outside to
the inside; look to see if last year's ring is narrow or
wide; a narrow ring means a dry year; a wide ring
means a rainy year.

Animals (pp 29–30)

A TRUE/FALSE

1. T   2. F, food   3. F, some   4. T   5. T

B MULTIPLE CHOICE

6. d   7. a   8. b   9. d   10. b

C COMPLETION

11. senses   12. arms   13. trees   14. eight
15. food

D SHORT ANSWERS

16. 100%   17. smaller   18. same   19. smaller
20. different

E WRITING

An owl is made of many cells. It eats food to get
energy. It moves around. Its senses help it respond to
things around it. Its senses help it find mice to eat
and a tree to live in. It reproduces when it makes
baby owls.

Invertebrates (pp 31–32)

A TRUE/FALSE

1. F, eight   2. T   3. F, don't have   4. T   5. T

B MULTIPLE CHOICE

6. d   7. a   8. d   9. c   10. b

C COMPLETION

11. closed   12. body   13. Water   14. chrysalis
15. tube

D SHORT ANSWERS

16. It hatches into a caterpillar.   17. It forms
a chrysalis.   18. The caterpillar changes into a
butterfly.   19. four   20. metamorphosis

E WRITING

When I become an adult, I stop swimming. I can't
move. I stay in one place. I pull in water through tiny
openings in my body called pores. I get food from the
water. The water leaves through the top opening in
my body.

Vertebrates (pp 33–34)

A TRUE/FALSE

1. F, warm-blooded   2. T   3. F, mammal   4. T   5. T

B MULTIPLE CHOICE

6. b   7. d   8. c   9. c   10. b

C COMPLETION

11. newborn   12. heat   13. temperature
14. surroundings   15. its mother's milk

GATEWAY TO SCIENCE Assessment Book • Copyright © Thomson Heinle

D SHORT ANSWERS

16. A mammal's stays the same but a fish's changes.   17. a reptile's   18. Their body temperatures change.   19. backbones
20. warm-blooded

E WRITING

When I was born I crawled across my mother's fur. I climbed into her pouch. I drank my mother's milk and got bigger. I lived in the pouch until I was nine months old.

The Human Body (pp 35–36)

A TRUE/FALSE

1. T   2. T   3. F, eleven   4. T   5. T

B MULTIPLE CHOICE

6. c   7. d   8. c   9. a   10. b

C COMPLETION

11. left   12. circular   13. Bones   14. circulatory
15. skin

D SHORT ANSWERS

16. the kidneys   17. to move gases in and out of the body   18. the lungs   19. the digestive system   20. the brain and the nerve cells

E WRITING

answers will vary; students may choose nervous system (brain and nerve cells control body's response to world and activities such as breathing); excretory system (kidneys control waste removal); respiratory system (lungs move gases in and out of body); digestive system (stomach and intestines change food into a form the body can use); circulatory system (heart and blood vessels take food and gases to every cell).

Asexual Reproduction (pp 37–38)

A TRUE/FALSE

1. T   2. T   3. F, stage   4. F, some living things   5. T

B MULTIPLE CHOICE

6. a   7. b   8. a   9. d   10. c

C COMPLETION

11. leaves   12. chromosome   13. runners
14. cells   15. mitosis

D SHORT ANSWERS

16. 2   17. 4   18. It doubles.   19. 16   20. 75

E WRITING

An onion can grow a new bulb. A new plant can grow from the bulb. The strawberry can grow runners.

Runners are stems that grow along the top of the soil. New plants can grow at special places where the runner touches the ground.

Sexual Reproduction (pp 39–40)

A TRUE/FALSE

1. F, identical to   2. T   3. T   4. T   5. F, two parents

B MULTIPLE CHOICE

6. d   7. c   8. a   9. d   10. c

C COMPLETION

11. sexual   12. crossing over   13. an egg
14. sexual reproduction   15. order

D SHORT ANSWERS

16. 2   17. sexual reproduction   18. 1
19. egg cell   20. offspring

E WRITING

The puppies have two parents; when sex cells from different parents join together, there are changes in the chromosomes; the puppies have variations because each puppy has some chromosomes from each parent.

Genetics (pp 41–42)

A TRUE/FALSE

1. F, DNA   2. F, genes   3. F, Genes control
4. F, genes   5. F, genes

B MULTIPLE CHOICE

6. b   7. a   8. c   9. d   10. b

C COMPLETION

11. nucleus   12. traits   13. instructions
14. recessive   15. ladder

D SHORT ANSWERS

16. small or lowercase   17. 1 dominant and 1 recessive   18. one   19. one   20. three

E WRITING

DNA looks like a twisted ladder. The "rungs" of the ladder are made of the bases adenine, thymine, guanine, and cytosine. DNA contains instructions on how to build each body part.

Changes Over Time (pp 43–44)

A TRUE/FALSE

1. F, can have   2. T   3. T   4. T   5. T

B MULTIPLE CHOICE

6. d   7. d   8. b   9. b   10. a

C COMPLETION

11. subspecies   12. species   13. natural selection
14. tree trunks   15. seeds

D SHORT ANSWERS

16. albatrosses   17. a gull   18. loons
19. Archaeopteryx   20. cuckoo

E WRITING

Over time, birds' beaks changed. They adapted to
help birds eat the food they found. Different beaks are
adaptations to eat different foods. Crossed beaks
help some crossbills get seeds from pinecones. Short,
strong beaks help some grosbeaks eat seeds. Long,
thin beaks help herons catch fish. Pouched beaks
help pelicans scoop fish from the water.

Classification Systems (pp 45–46)

A TRUE/FALSE

1. F, two names   2. T   3. T   4. F, wet places   5. T

B MULTIPLE CHOICE

6. c   7. d   8. b   9. a   10. a

C COMPLETION

11. species name   12. fungi   13. stomachs
14. genus   15. kingdoms

D SHORT ANSWERS

16. mammal   17. cheetah, tiger, and lion
18. carnivore   19. fly and bird   20. fly

E WRITING

Carolus Linnaeus was a scientist. He created a
classification system. It is called "binomial nomenclature."
The system gives each species a two-part name. It
tells us the genus name and the species name.

Biomes and Ecosystems (pp 47–48)

A TRUE/FALSE

1. F, biome   2. T   3. T   4. F, wet   5. T

B MULTIPLE CHOICE

6. c   7. b   8. d   9. a   10. c

C COMPLETION

11. lose their leaves   12. water   13. timeline
14. desert   15. species

D SHORT ANSWERS

16. in 1900   17. grasses   18. 1930   19. 1990
20. 120 years

E WRITING

A kangaroo rat lives in a hot and dry habitat. The
kangaroo rat lives in a hole in the ground. That burrow
is its home. The kangaroo rat eats seeds. It gets
water from the food it eats.

Energy Transfer in Living Things
(pp 49–50)

A TRUE/FALSE

1. T   2. F, consumers   3. T   4. F, predator   5. T

B MULTIPLE CHOICE

6. b   7. a   8. d   9. d   10. a

C COMPLETION

11. helped   12. hide   13. good web
14. consumer   15. moves

D SHORT ANSWERS

16. light   17. consumers   18. heat   19. the top
20. some energy is used for growth at each level

E WRITING

A snake is a predator that gets energy by catching
and eating other animals; those animals are its prey;
those animals may get energy by eating plants or
smaller animals; possible food chain might be: grass,
grasshopper, frog, snake, owl.

Cycles in Nature (pp 51–52)

A TRUE/FALSE

1. T   2. F, carbon dioxide   3. T   4. F, roots
5. F, oxygen

B MULTIPLE CHOICE

6. c   7. a   8. d   9. a   10. c

C COMPLETION

11. the soil   12. carbon dioxide   13. fall
14. nitrogen   15. groundwater

D SHORT ANSWERS

16. Bacteria breaks down animal wastes.
17. nitrogen   18. from bacteria in the soil
19. the soil and the atmosphere   20. where
nitrogen travels

E WRITING

The plant gives off oxygen for me to breathe. I breathe
out carbon dioxide that the plant uses. I also feed and
water the plant. Someday I will eat the carrot for food.

Responding to the Environment
(pp 53–54)

A TRUE/FALSE

1. F, stimulus   2. F, in spring   3. T   4. T   5. F, reflex

B MULTIPLE CHOICE

6. c   7. d   8. b   9. d   10. c

GATEWAY TO SCIENCE Assessment Book • Copyright © Thomson Heinle

C COMPLETION

11. flowers   12. learned behavior   13. stimulus
14. insects   15. hibernates

D SHORT ANSWERS

16. closing around the insect   17. food
18. It will open again.   19. gravity
20. turn to the light

E WRITING

A chipmunk hibernates in the winter. It sleeps during
the cold weather. A frog estivates in the summer. It
sleeps during the hot weather. Both animals are in a
deep sleep to live through the weather.

Conservation (pp 55–56)

A TRUE/FALSE

1. F, extinct   2. T   3. T   4. F, are not extinct   5. T

B MULTIPLE CHOICE

6. d   7. c   8. a   9. b   10. c

C COMPLETION

11. fish   12. rain forest   13. humans or people
14. break   15. species

D SHORT ANSWERS

16. extinct   17. about 4,000   18. Sumatran
tigers   19. South China tiger   20. 3

E WRITING

People use plants as medicines. Scientists think
other rain forest plants can be useful. Scientists
think they can use the plants to make new medicines.

Life Science End-of-Section Test
(pp 57–62)

A TRUE/FALSE

1. T   2. T   3. F, seed   4. T   5. F, warm blooded
6. T   7. F, DNA   8. F, two names   9. T
10. F, stimulus

B MULTIPLE CHOICE

11. a   12. a   13. c   14. b   15. d   16. d
17. d   18. d   19. b   20. c   21. a   22. d
23. b   24. d   25. c   26. c   27. b   28. c
29. c   30. d

C COMPLETION

31. fungus   32. nutrients   33. rain   34. closed
35. left   36. sexual   37. subspecies   38. leaves
39. soil   40. fish

D SHORT ANSWERS

41. mammal   42. cheetah, tiger, and lion
43. carnivore   44. fly and bird   45. fly   46. 2
47. 4   48. It doubles.   49. 16   50. 75

E WRITING ASSESSMENT

51. Both are seedless plants; both reproduce by
making spores; ferns are vascular plants, mosses are
nonvascular plants; ferns have roots, but mosses do
not; ferns also have xylem and phloem, but mosses
do not.

52. I leave the left side of the heart. I go through
the body in blood vessels. Then I go back to the right
side of the heart.

53. Members of species have the same adaptations.
Members of one species cannot reproduce with
members of another species. Herons and pelicans are
members of different species. Members of a subspecies
have some variations. Members of subspecies can
reproduce together. Different kind of seaside sparrows
are members of different subspecies.

54. The 1950s; people polluted waterways with DDT;
eagles ate polluted fish; eagles' eggshells became
weak; their eggs broke; many baby eagles died.

Space (pp 63–64)

A TRUE/FALSE

1. F, billions   2. F, stars   3. T   4. F, star   5. F, big

B MULTIPLE CHOICE

6. c   7. a   8. c   9. c   10. d

C COMPLETION

11. radio waves   12. in one year   13. AU or
astronomical unit   14. mirrors   15. kilometers

D SHORT ANSWERS

16. Venus   17. 0.5 AU   18. 9.5 AU   19. Uranus
20. Mercury

E WRITING

Telescopes collect light waves with lenses or with
lenses and mirrors. Radio telescopes collect radio
waves in bowl-shaped dishes. Both telescopes and
radio telescopes make things that are far away look
larger. They help us see farther into space.

Stars (pp 65–66)

A TRUE/FALSE

1. F, brightness   2. T   3. T   4. F, yellow   5. F, small

B MULTIPLE CHOICE

6. c   7. d   8. b   9. a   10. c

C COMPLETION

11. nebula   12. gravity   13. dust and gas
14. close   15. constellations

D SHORT ANSWERS

16. hotter   17. dim   18. hotter   19. dimmer
20. on the lower right

E WRITING

A main sequence star is born in a nebula. It burns hydrogen as fuel. When it starts to burn all its fuel, it swells into a red giant. Then it shrinks and becomes a white dwarf. As it shrinks, it leaves behind a new nebula.

Our Solar System (pp 67–68)

A TRUE/FALSE

1. F, comet   2. T   3. T   4. F, Jupiter
5. F, stretched-out orbits

B MULTIPLE CHOICE

6. b   7. d   8. c   9. b   10. a

C COMPLETION

11. comet   12. moons   13. the sun
14. the sun   15. Halley's comet

D SHORT ANSWERS

16. four   17. Callisto   18. Callisto
19. the Great Red Spot   20. larger

E WRITING

Halley's commet is a very bright comet. It appears every 76 years. It is made of ice, rock, and gases. The heat from the sun melts some of the ice. The ice makes a glowing tail. The comet is named after Sir Edmund Halley because he figured out that comets seen years apart were really the same object.

Earth, the Moon, and the Sun (pp 69–70)

A TRUE/FALSE

1. T   2. T   3. T   4. T   5. F, day

B MULTIPLE CHOICE

6. c   7. a   8. b   9. b   10. d

C COMPLETION

11. seasons   12. southern half   13. shape
14. away from the sun   15. the sun

D SHORT ANSWERS

16. Nome, Alaska   17. Nome, Alaska   18. about 6 hours   19. Miami, Florida   20. Chicago

E WRITING

Earth rotates on its axis. This takes 24 hours. It also revolves around the sun in an orbit. It takes about 365 days for Earth to revolve around the sun.

Eclipses and Tides (pp 71–72)

A TRUE/FALSE

1. T   2. F, moon   3. T   4. T   5. F, twice

B MULTIPLE CHOICE

6. b   7. a   8. d   9. c   10. b

C COMPLETION

11. tidal range   12. partial solar   13. meters
14. line up   15. right angle

D SHORT ANSWERS

16. 12:41 P.M.   17. 6:36 P.M.   18. low tide
19. around midnight or 12:41 A.M.   20. It was too dark to take a photo

E WRITING

Earth is between the sun and the moon. Earth casts a shadow on the moon. The moon's light is blocked. The moon is completely covered in shadow.

Space Exploration (pp 73–74)

A TRUE/FALSE

1. T   2. F, telescopes   3. T   4. T   5. F, can

B MULTIPLE CHOICE

6. d   7. d   8. c   9. a   10. b

C COMPLETION

11. the moon   12. the International Space Station or ISS   13. stars and planets
14. space suits   15. satellite

D SHORT ANSWERS

16. in 1957   17. 4 years   18. Apollo 11 astronauts walked on the moon.   19. in 1981
20. Building of the space station began.

E WRITING

Activities should include: eat, do experiments, exercise, make repairs, and sleep. Students may assign these activities to any time of day. Scientists learn how humans can live and work in space.

Minerals and Rocks (pp 75–76)

A TRUE/FALSE

1. F, sedimentary   2. F, mineral   3. T
4. F, form   5. T

B MULTIPLE CHOICE

6. b   7. a   8. d   9. b   10. c

C COMPLETION

11. minerals   12. sedimentary   13. change
14. 3   15. sedimentary

GATEWAY TO SCIENCE Assessment Book • Copyright © Thomson Heinle

D SHORT ANSWERS

16. heat and pressure   17. It becomes sedimentary rock.   18. igneous rock and metamorphic rock   19. igneous rock   20. igneous, metamorphic, and sedimentary rocks

E WRITING

It is probably sedimentary rock. Sedimentary rock is often found near rivers and oceans. The plant was pressed together in layers with bits of sand, shell, and rock. After many years it became part of the rock.

Earth's Structure (pp 77–78)

A TRUE/FALSE

1. F, plates   2. T   3. T   4. T   5. T

B MULTIPLE CHOICE

6. c   7. c   8. b   9. a   10. a

C COMPLETION

11. Pangaea   12. seafloor spreading   13. upper mantle   14. liquid metal   15. continental drift

D SHORT ANSWERS

16. South America   17. south   18. Asia and Europe   19. North America and Africa   20. north

E WRITING

It helps explain continental drift. It tells the scientist that the same kinds of plants and animals lived in Africa and South America. Those continents were once one continent. Fossils formed. The continents later drifted apart.

Earth's Surface (pp 79–80)

A TRUE/FALSE

1. T   2. F, towards   3. F, Asia   4. F, triangular   5. T

B MULTIPLE CHOICE

6. c   7. b   8. a   9. d   10. c

C COMPLETION

11. Tibet   12. water   13. glacier   14. build up   15. shape

D SHORT ANSWERS

16. about 32 meters   17. about 22 meters   18. about 12 meters   19. about 10 meters   20. 0 meters

E WRITING

The Nile River flows toward the ocean; it carries soil and rock; when the river meets the ocean, water slows down; soil and rock drop to the bottom; over time, soil and rock build up into a delta.

Earthquakes and Volcanoes (pp 81–82)

A TRUE/FALSE

1. T   2. T   3. F, the crust   4. T   5. F, at the boundaries

B MULTIPLE CHOICE

6. a   7. a   8. c   9. d   10. b

C COMPLETION

11. together   12. epicenter   13. hot spots   14. fault   15. pressure

D SHORT ANSWERS

16. a convergent boundary   17. past each other   18. arrows   19. drawing A   20. drawing A

E WRITING

Earth's plates pull apart or come together. The rock at the boundaries gets very hot and becomes magma. Magma rises up and forms a volcano. The Hawaiian Islands are over a hot spot, a place where the mantle is very hot. Volcanoes often form at hot spots.

Our Changing Earth (pp 83–84)

A TRUE/FALSE

1. T   2. T   3. T   4. F, slowly   5. T

B MULTIPLE CHOICE

6. d   7. b   8. b   9. d   10. a

C COMPLETION

11. waves   12. crack   13. glacier   14. Wind   15. volcanoes

D SHORT ANSWERS

16. sand dune   17. the plant root   18. water erosion   19. chemical weathering   20. mechanical weathering or abrasion

E WRITING

The house will weather. The wind will blow tiny bits of rock against the house. Abrasion will break down the stone. This is mechanical weathering. Chemicals in rain water will react with chemicals in the house. The stone in the house will break down over time. This is chemical weathering.

The Atmosphere (pp 85–86)

A TRUE/FALSE

1. T   2. F, small   3. T   4. F, outside   5. F, several types

B MULTIPLE CHOICE

6. a   7. c   8. b   9. c   10. c

C COMPLETION

11. nitrogen   12. gases   13. oxygen
14. trace gases   15. troposphere

D SHORT ANSWERS

16. 78%   17. 21%   18. 1%   19. carbon
dioxide and water vapor   20. the one showing
trace gases

E WRITING

troposphere; the temperature is warm; in other layers the
air is too cold and thin for plants and animals to live.

Weather and Climate (pp 87–88)

A TRUE/FALSE

1. T   2. T   3. F, polar   4. T   5. F, slowly

B MULTIPLE CHOICE

6. b   7. b   8. c   9. d   10. b

C COMPLETION

11. precipitation   12. average   13. cloud
14. weight   15. climate

D SHORT ANSWERS

16. low   17. rainy   18. south   19. low
20. cold and stationary

E WRITING

Weather means the conditions at a certain time. It can
change in an hour. Climate is the average weather
over many years. It changes slowly.

Extreme Weather (pp 89–90)

A TRUE/FALSE

1. F, warming   2. T   3. T   4. F, hurricanes   5. T

B MULTIPLE CHOICE

6. b   7. d   8. c   9. a   10. d

C COMPLETION

11. tropical   12. name   13. rain or hail
14. warnings   15. quickly

D SHORT ANSWERS

16. a cumulonimbus cloud   17. a wall cloud
18. above   19. rain   20. a tornado

E WRITING

Warm, moist air rose into the atmosphere in an updraft.
It met cold air. The moisture in the warm air fell as
rain or hail. The rain or hail fell in a downdraft.

Natural Resources (pp 91–92)

A TRUE/FALSE

1. T   2. F, renewable   3. T   4. T   5. F, minerals

B MULTIPLE CHOICE

6. c   7. d   8. b   9. c   10. a

C COMPLETION

11. fossil   12. water   13. wind machine
14. natural resources   15. fuel

D SHORT ANSWERS

16. 8%   17. 34%   18. 92%   19. 17%
20. 85%

E WRITING

Coal, oil, petroleum, and natural gas are fossil fuels.
They are nonrenewable because they take millions of
years to form. They cannot be replaced in our lifetimes.
When they are used, they can't be replaced quickly.

Earth Science End-of-Section Test
(pp 93–98)

A TRUE/FALSE

1. F, billions   2. F, comet   3. T   4. F, sedimentary
5. T   6. T   7. T   8. T   9. F, mineral   10. T

B MULTIPLE CHOICE

11. c   12. c   13. b   14. c   15. b   16. d
17. b   18. c   19. c   20. a   21. d   22. a
23. b   24. b   25. c   26. a   27. d   28. d
29. a   30. b

C COMPLETION

31. nebula   32. seasons   33. the moon
34. Pangaea   35. together   36. nitrogen
37. tropical   38. seafloor spreading   39. gases
40. partial solar

D SHORT ANSWERS

41. about 30 meters   42. about 20 meters
43. about 10 meters   44. about 10 meters
45. 0 meters   46. in 1957   47. 4 years
48. Apollo 11 astronauts walked on the moon.
49. in 1981   50. Building of the space station
began.

E WRITING ASSESSMENT

51. The northern half of Earth is tilted toward the
sun. It is warm there. The days are long.

52. The sediment is pressed into sedimentary rock.
Heat and pressure change the sedimentary rock to
metamorphic rock. The metamorphic rock melts into
magma. It then cools into igneous rock.

53. Water can get into a crack in a rock. The water
freezes into ice and expands. The ice makes the crack
get bigger. Finally, the rock breaks in two. Glaciers can
wear down mountains and change their shape. Glaciers
can form moraines.

54. Winds in the disturbance get faster. It becomes a tropical depression with rotating wind; wind speeds are less than 62 km/h. Winds get even faster, more than 62 km/h. It becomes a tropical storm. When winds reach 117 km/h, it becomes a hurricane with an eye, lots of rain, and storm surges.

Nature of Matter (pp 99–100)

A TRUE/FALSE
1. F, three   2. F, volume   3. T   4. F, gas   5. F, solid

B MULTIPLE CHOICE
6. c   7. c   8. b   9. d   10. c

C COMPLETION
11. sight   12. physical   13. physical properties
14. chemical properties   15. senses

D SHORT ANSWERS
16. physical change   17. solid   18. picture A
19. liquid and gas   20. gas

E WRITING
Answers will vary; students can mention using sight to observe color, using smell to observe odor, using taste to observe taste or sweetness, using touch to observe texture, using sight or touch to observe size and shape, and so on.

Measuring Matter (pp 101–102)

A TRUE/FALSE
1. T   2. F, volume   3. F, boils   4. T   5. T

B MULTIPLE CHOICE
6. c   7. b   8. c   9. b   10. a

C COMPLETION
11. boiling point   12. density   13. down
14. weight   15. floats

D SHORT ANSWERS
16. oxygen   17. salt   18. 1,313°C   19. oxygen
20. 612°C

E WRITING
You weigh less on the moon. The moon has less gravity than Earth. Weight is caused by gravity. Your mass stays the same. Mass is the amount of matter in an object. It doesn't change because of gravity.

Atoms and Molecules (pp 103–104)

A TRUE/FALSE
1. T   2. F, atoms   3. T   4. F, atoms   5. F, metal

B MULTIPLE CHOICE
6. b   7. c   8. c   9. a   10. d

C COMPLETION
11. positive   12. liquid   13. combine
14. nonmetals   15. matter

D SHORT ANSWERS
16. the parts of an atom   17. protons and neutrons   18. electrons   19. 2   20. 2

E WRITING
Protons and neutrons are in the nucleus or center of the atom; electrons circle the nucleus; protons and neutrons are larger than electrons; protons have a positive charge; electrons have a negative charge; neutrons have no charge.

Compounds and Mixtures (pp 105–106)

A TRUE/FALSE
1. F, chemical change   2. T   3. T   4. T
5. F, an element or a compound

B MULTIPLE CHOICE
6. d   7. d   8. b   9. b   10. b

C COMPLETION
11. physical   12. solution   13. physical
14. not the same   15. table salt

D SHORT ANSWERS
16. water and table salt   17. sugar, baking soda, and chalk   18. baking soda   19. table salt
20. oxygen and hydrogen

E WRITING
You start with sugar and water. Sugar and water are compounds. Sugar is made of carbon, hydrogen, and oxygen. Water is made of hydrogen and oxygen. When you mix them you get a solution. A solution is a homogeneous mixture. This is a physical change.

Chemical Reactions (pp 107–108)

A TRUE/FALSE
1. F, exothermic   2. T   3. T   4. T   5. F, cannot be

B MULTIPLE CHOICE
6. b   7. d   8. a   9. d   10. c

C COMPLETION
11. reactions   12. product   13. symbol
14. atoms   15. equal

D SHORT ANSWERS
16. 4   17. 3   18. 2   19. that there are two atoms in the molecule   20. the direction of the reaction

E WRITING

Two atoms of sodium join with two atoms of chlorine. They form two molecules of salt. A molecule of salt has one atom of sodium and one atom of chlorine.

Radiation and Radioactivity
(pp 109–110)

A TRUE/FALSE

1. T   2. F, energy   3. T   4. T   5. F, radioactivity

B MULTIPLE CHOICE

6. a   7. b   8. c   9. a   10. c

C COMPLETION

11. energy   12. instruments   13. radioactive
14. particles   15. radioactive

D SHORT ANSWERS

16. carbon-14   17. uranium-238   18. in nuclear power plants   19. to clean surgical instruments
20. 5,730 years

E WRITING

We remember Antoine-Henri Becquerel because he discovered radioactivity. He left a photographic plate in a drawer with some uranium. In a few days there was an image of the uranium on the plate. His discovery was important because it helped us understand radiation.

Forces (pp 111–112)

A TRUE/FALSE

1. T   2. F, unbalanced   3. T   4. T   5. F, force

B MULTIPLE CHOICE

6. c   7. d   8. a   9. c   10. c

C COMPLETION

11. mass   12. attraction   13. faster
14. gravity   15. equilibrium

D SHORT ANSWERS

16. Mercury   17. Saturn   18. Uranus
19. more   20. Mars

E WRITING

The flag is not moving because the forces are balanced. Each side is pulling the same amount. When the teacher joins one side, the forces become unbalanced. One side is pulling harder. The flag moves.

Forces and Motion (pp 113–114)

A TRUE/FALSE

1. T   2. F, speed   3. F, force   4. T   5. F, three

B MULTIPLE CHOICE

6. d   7. c   8. b   9. d   10. c

C COMPLETION

11. acceleration   12. moving forward
13. velocity   14. force   15. speed

D SHORT ANSWERS

16. 20 newtons   17. 5 m/s/s   18. 7 kg
19. The acceleration doubles.   20. You get half the acceleration.

E WRITING

When the batter hits the ball an action force pushes against the ball. A reaction force pushes against the bat. When the fielder catches the ball, an action force pushes against the glove. A reaction force pushes against the ball.

Work, Power, and Machines
(pp 115–116)

A TRUE/FALSE

1. F, easier   2. T   3. T   4. T   5. T

B MULTIPLE CHOICE

6. d   7. a   8. c   9. c   10. d

C COMPLETION

11. downward   12. direction   13. simple machine   14. simple machine   15. compound

D SHORT ANSWERS

16. person B   17. no   18. decreases it
19. It stays the same.   20. more

E WRITING

Wheeling in a wheelbarrow will take less force; the wheel and axle change the direction of the force, so it's easier to push the wheelbarrow than to carry a basket. The amount of work stays the same.

Waves (pp 117–118)

A TRUE/FALSE

1. F, low   2. F, longitudinal   3. T   4. T   5. T

B MULTIPLE CHOICE

6. c   7. a   8. d   9. a   10. c

C COMPLETION

11. wavelength   12. X-rays   13. longitudinal
14. amplitude   15. radio

D SHORT ANSWERS

16. longitudinal wave   17. transverse wave
18. 1. (transverse wave)   19. 2. (transverse wave)
20. 1. (longitudinal wave)

GATEWAY TO SCIENCE Assessment Book • Copyright © Thomson Heinle

1

E WRITING
possibilities include light waves to see, sound waves to hear, radio waves to receive radio or TV signals, microwaves to cook.

Light (pp 119–120)

A TRUE/FALSE
1. F, transparent   2. F, reflect   3. T   4. T
5. F, the opposite thing

B MULTIPLE CHOICE
6. d   7. c   8. a   9. b   10. c

C COMPLETION
11. colors   12. prism   13. refraction   14. color
15. opaque

D SHORT ANSWERS
16. smooth, shiny   17. It is refracted.
18. a transparent object   19. It is absorbed.
20. an opaque object

E WRITING
The bike looks like a certain color because the paint reflects that color light and absorbs other colors. The headlight and taillight have transparent or translucent covers. Shiny parts made of metal or chrome reflect more light than other parts.

Forms of Energy (pp 121–122)

A TRUE/FALSE
1. T   2. F, electrical   3. T   4. F, light energy
5. F, nonrenewable

B MULTIPLE CHOICE
6. b   7. d   8. b   9. c   10. b

C COMPLETION
11. chemical   12. thermal   13. kinetic
14. sound   15. renewable

D SHORT ANSWERS
16. at the top of the hill   17. while going down the hill   18. while going down the hill
19. mechanical energy   20. mechanical energy

E WRITING
light energy from the candle, thermal energy from the fire, sound energy from the CD player, electrical energy to run the refrigerator, chemical energy to run the CD player

Energy Transformations (pp 123–124)

A TRUE/FALSE
1. F, temperature   2. T   3. F, conduction
4. F, hot   5. T

B MULTIPLE CHOICE
6. c   7. d   8. a   9. c   10. b

C COMPLETION
11. potential energy   12. heat   13. sound energy
14. energy   15. across

D SHORT ANSWERS
16. electrical energy   17. thermal energy
18. by conduction   19. the water at the bottom of the pan   20. by convection

E WRITING
electrical energy in the toaster changes to thermal energy; thermal energy moves to the bread by radiation; electrical energy or energy from gas in the stove changes to thermal energy; thermal energy moves from the pan to the egg by convection.

Energy and Life (pp 125–126)

A TRUE/FALSE
1. T   2. F, sugar   3. T   4. T   5. T

B MULTIPLE CHOICE
6. a   7. c   8. d   9. c   10. b

C COMPLETION
11. cellular   12. chemical   13. oxygen
14. food   15. chlorophyll

D SHORT ANSWERS
16. the sun   17. the atmosphere   18. the ground   19. the plant   20. chemical energy

E WRITING
the leaves on the apple tree trapped energy in photosynthesis; chlorophyll in the leaves used sunlight, water, and carbon dioxide to make glucose; the tree stores the glucose in the apple

Electricity and Magnetism (pp 127–128)

A TRUE/FALSE
1. T   2. T   3. T   4. T   5. T

B MULTIPLE CHOICE
6. d   7. a   8. c   9. c   10. a

C COMPLETION
11. the south pole   12. magnetic field
13. electric generator   14. static   15. electrons

D SHORT ANSWERS
16. a lightbulb   17. a battery   18. a switch
19. a wire   20. The circuit is broken because the switch is open.

E WRITING

static electricity; one object loses electrons; one object gains electrons; the electrons jump from one place to another

Physical Science End-of-Section Test
(pp 129–134)

A TRUE/FALSE

1. F, 3   2. T   3. F, exothermic   4. T
5. F, easier   6. F, transparent   7. F, temperature
8. T   9. F, volume   10. F, volume

B MULTIPLE CHOICE

11. c   12. c   13. b   14. d   15. b   16. a
17. c   18. d   19. d   20. c   21. d   22. b
23. c   24. a   25. d   26. c   27. d   28. d
29. b   30. d

C COMPLETION

31. boiling point   32. physical   33. energy
34. acceleration   35. wavelength   36. chemical
37. cellular   38. X-rays   39. thermal
40. chemical

D SHORT ANSWERS

41. 20 newtons   42. 5 m/s/s   43. 7 kg
44. The acceleration doubles.   45. You get half the acceleration.   46. electrical energy
47. thermal energy   48. by conduction
49. the water at the bottom of the pan
50. by convection

E WRITING ASSESSMENT

51. A solid has particles that are tightly packed together. It has a definite shape and volume. The particles are farther apart in a liquid. A liquid has a definite volume. It takes the shape of its container. The particles in a gas are very far apart. Water vapor fills its container.

52. A seesaw is a board that turns on a fulcrum; it makes work easier because it changes the direction of force; instead of using upward force, you use the lever and a downward force.

53. When the light enters the raindrops it refracts. Each color light bends differently. The light passes through the raindrops and separates into the colors of the rainbow.

54. To make a magnet, wrap an iron bar with wire, run electricity through the wire, and the bar becomes a magnet. To make an electric current, spin metal wires inside a magnetic field.

END-OF-BOOK-TEST (pp 135–140)

A TRUE/FALSE

1. F, two circles   2. F, many kinds   3. F, seeds
4. T   5. F, stars   6. T   7. T   8. T   9. T   10. T

B MULTIPLE CHOICE

11. a   12. c   13. a   14. a   15. d   16. d
17. b   18. c   19. a   20. a   21. a   22. b
23. d   24. c   25. c   26. a   27. c   28. d
29. a   30. b

C COMPLETION

31. conclusion   32. species name   33. helped
34. flowers   35. the International Space Station or ISS   36. oxygen   37. precipitation   38. oxygen
39. magnetic field   40. electric generator

D SHORT ANSWERS

41. 212 degrees   42. 100 degrees   43. 32 degrees
44. Fahrenheit scale   45. 20 degrees   46. 8%
47. 34%   48. 92%   49. 17%   50. 85%

E WRITING ASSESSMENT

51. They should wear goggles to protect their eyes; they should wear lab aprons to protect their clothing and bodies; they should wear gloves to protect their hands; they should wear heatproof gloves to protect their hands when working with hot objects.

52. Plant and animal cells have many of the same parts: cell membrane, cytoplasm, nucleus, vacuoles, ribosomes, and other organelles; a plant cell has chloroplasts and a cell wall; animal cells do not.

53. A supergiant is a very large star. It is many times larger than the sun. It can explode and become a supernova. Its life can end as a neutron star.

54. Water is made of hydrogen and oxygen; it has two hydrogen atoms and one oxygen atom; hydrogen and oxygen at room temperature are gases; water at room temperature is a liquid.

GATEWAY TO SCIENCE Assessment Book • Copyright © Thomson Heinle